T0191873

Analog Circuits and Signal Processing

More information about this series at http://www.springer.com/series/7381

Pedro Emiliano Paro Filho • Jan Craninckx
Piet Wambacq • Mark Ingels

Charge-based CMOS Digital RF Transmitters

Pedro Emiliano Paro Filho
SSET department
IMEC SSET department
Leuven, Belgium

Jan Craninckx
SSET department
IMEC SSET department
Leuven, Belgium

Piet Wambacq
SSET department
IMEC SSET department
Leuven, Belgium

Mark Ingels
SSET department
IMEC SSET department
Leuven, Belgium

ISSN 1872-082X ISSN 2197-1854 (electronic)
Analog Circuits and Signal Processing
ISBN 978-3-319-83372-9 ISBN 978-3-319-45787-1 (eBook)
DOI 10.1007/978-3-319-45787-1

Printed on acid-free paper

This Springer imprint is published by Springer Nature
The registered company is Springer International Publishing AG
The registered company address is: Gewerbestrasse 11, 6330 Cham, Switzerland

If Education alone cannot transform society,
nor without it society changes.

Paulo Freire

To my family

the leaning,
the standing, and
the crawling ones.

Preface

The point where wireless communication and ubiquitous connectivity became an essential part of our lives is already past. Generation after generation communication speed is being taken to unprecedented levels, requiring both state-of-the-art hardware and software to handle a huge volume of data, delivered to an increasing number of users in an overcrowded spectrum. To our delight, the challenges are always plentiful.

With respect to the radio front-end, providing an extremely low noise emission with an improved signal integrity is a key requirement to support high-order modulation schemes (e.g., 64 QAM) in situations where anyone's transmitter can be interfering with a neighbor user or its own receiver in frequency-division full duplex mode. Increasing power and/or area consumption is not an option in this case. On the contrary, for an improved user experience the battery should last longer, and the price per component should always go down, so that more and more features can be added to a mobile/handheld device. Thus, making a better performing CMOS radio front-end that consumes even less power and area is a hot research topic these days, especially regarding the transmitter and PA designs, considered by many the "battery killers" on most mobile devices.

A quick analysis of literature shows that with regard to CMOS transmitter implementations, the state of the art is clearly divided into analog- and digital-intensive architectures. In terms of out-of-band noise, analog-intensive architectures are undoubtedly the best performing implementations. However, their improved noise performance is typically achieved through extensive low-pass filtering along the entire signal path, which has a significant impact on area consumption. Digital-intensive implementations, on the other hand, are by far the most portable, area-efficient, and scaling-friendly ones. However, the lack of filtering (for both noise and aliases) makes it very challenging to meet the stringent out-of-band noise requirements in SAW-less operation.

In this book, a novel digital-intensive transmitter architecture that can relax this trade-off is described. Through the combination of charge-domain operation with incremental signaling, this architecture gives the best of both worlds, providing the reduced area and high portability of digital-intensive architectures with an improved out-of-band noise performance given by intrinsic noise filtering capabilities.

Two implementations of the incremental charge-based TX are demonstrated, differing on how the charge-based DAC (QDAC) is implemented and the RF load being driven: In the first realization, the RF load is the input capacitance of a PPA stage, and the QDAC is implemented with a controllable capacitance that is alternately pre-charged and connected to a charge reservoir. In the second implementation, the ability of delivering more power using the charge-based architecture has been investigated with a direct-launch architecture, where the 50 Ω load representing the PA input is directly driven with charge. The QDAC is implemented with a 12-bit conductance array, which proves to be the most area-efficient implementation in this case.

Prototyped using a 28 nm 0.9 V CMOS technology, both charge-based TX realizations provide remarkable results in terms of noise performance, thanks to their intrinsic noise filtering capability, improved sampling alias attenuation, and reduced quantization noise. With an out-of-band noise spectral density of −159 dBc/Hz and a core area of 0.22 mm^2, the second implementation achieves—to the author's knowledge—the best out-of-band noise performance versus area consumption when compared to other similar works. ACLR and EVM performance are also among the best. As a result, this work paves the way for compact CMOS SAW-less transmitter implementations enabling advanced wireless communication systems, including 3G, 4G, and beyond.

Leuven, Belgium

Pedro Emiliano Paro Filho
Jan Craninckx
Piet Wambacq
Mark Ingels

Acknowledgments

At the age of 16, I found myself with a book in hands, which tells the story of a seagull named Jonathan Livingston Seagull ("Fernão Capelo Gaivota"). Jonathan Seagull had a great drive to learn and improve his flying skills. His passion took him to new places and ever higher altitudes, even setting him apart from his loved ones eventually. Flying was an obsession for him, only shadowed by his willingness to share what he has learned along this long—and many times lonesome—journey toward greater understanding and self-awareness.

Four years ago, another seagull started a journey that in many ways resembles the story of Jonathan Seagull, reason why this book came to my mind 16 years later. These were years of hard work and great dedication to learning, teaching, and creating. There were also various high-speed dives that turned out as great crashes against the water, leaving the seagull adrift many times. But more than anything, the lessons learned—eventually leading to few successful flights over beautiful landscapes—and the people met along the way made every piece of this journey worthwhile. For that, I'd like to thank:

Jan, Piet, and Mark, whose guidance, trust, and support not only made this journey possible but also served me as inspiration, teaching with lifelong dedication the values of serious and honest research.

All my colleagues and friends from IMEC, with whom I had the chance to share fruitful moments over coffee and beer, and the brazilian community, which was a great source of support during the last 4 years with great barbecues, feijoadas, anniversaries, and, unfortunately, farewell parties.

Along the way, there were also people that somehow made all the difference, and for whom I keep a special place in my heart: Adrian, a living proof that companionship and intelligence cannot be measured in kilograms; Karlinha, Cadu, and Guto, three great-hearted people always ready to offer a friendly shoulder; and Oscar, Ricardão, André, Marcão, Cássio, Gigi, João, and Mihnea, great friends whose discussions were always a good way to wash away microelectronics from an overloaded mind.

Thank you all for the great time.

Also, none of this would have been possible if it were not for the support and affection received from abroad. Daily phone calls and countless countdowns to our next encounter were the only way to keep us together at distance but never apart. And we made it! Te amo linda.

Finally, my greatest gratitude to my beloved family. My dear father and mother, to whom I dedicate this work. My siblings Paula, Roberta, and Renato, my brother-in-law Rodrigo, and my two "not so little anymore" princesses Ana Luiza and Laurinha.
You are the sunshine of my life.

How much more there is now to living! Instead of our drab slogging forth and back to the fishing boats, there's a reason to life! We can lift ourselves out of ignorance, we can find ourselves as creatures of excellence and intelligence and skill. We can be free!
We can learn to fly!

(Jonathan Livingston Seagull, by Richard Bach)

Contents

List of Figures

List of Tables

List of Abbreviations

ACLR	Adjacent-Channel Leakage Ratio
ACPR	Adjacent-Channel Power Ratio
AMPS	Advanced Mobile Phone Service
AM	Amplitude Modulation
BER	Bit-Error Rate
BW	Bandwidth
CA	Carrier Aggregation
CIM3	Third-order Counter Intermodulation
CIM5	Fifth-order Counter Intermodulation
CMOS	Complementary Metal Oxide Semiconductor
CORDIC	Coordinate Rotation for Digital Computer
CQDAC	Capacitive Charge-based Digital-to-Analog Converter
CS	Common Source
DAC	Digital-to-Analog Converter
DNL	Differential Nonlinearity
DPD	Digital Pre-Distortion
EDGE	Enhanced Data-rates for GSM Evolution
ET	Envelope Tracking
EVM	Error-Vector Magnitude
FDD	Frequency-Division Duplexing
FDMA	Frequency-Division Multiple Access
FM	Frequency Modulation
GBW	Gain-Bandwidth Product
GMSK	Gaussian Minimum Shift Keying
GPRS	General Packet Radio Service
GPS	Global Positioning System
GSM	Global System for Mobile communication
HD3	Third-order Harmonic Distortion
HD5	Fifth-order Harmonic Distortion
HD	Harmonic Distortion
HSDPA	High-Speed Downlink Packet Access

INL	Integral Nonlinearity
LO	Local Oscillator
LSB	Least-Significant Bit
LTE	Long Term Evolution
MIMO	Multiple Input Multiple Output
MOM	Metal-Oxide-Metal
MOS	Metal Oxide Semiconductor
NFC	Near Field Communication
NMOS	N-type Metal Oxide Semiconductor
NMT	Nordic Mobile Telephone
NOC	Network-On-Chip
NTT	Nippon Telegraph and Telephone
OFDM	Orthogonal Frequency-Division Multiplexing
OOB	Out-Of-Band (noise)
OQPSK	Offset Quadrature Phase-Shift Keying
OSR	Oversampling Ratio
OTA	Operational Transconductance Amplifier
PAPR	Peak-to-Average Power Ratio
PA	Power Amplifier
PCB	Printed Circuit Board
PMOS	P-type Metal Oxide Semiconductor
PM	Phase Modulation
PNOISE	Periodic Steady-State Noise
PPA	Pre-Power Amplifier
QAM	Quadrature Amplitude Modulation
QDAC	Charge-based Digital-to-Analog Converter
RAM	Random Access Memory
RFDAC	Typical denomination for direct Digital-to-RF Converter
RF	Radio Frequency
RMS	Root Mean Square
RQDAC	Resistive Charge-based Digital-to-Analog Converter
SAW	Surface Acoustic Wave
SC-R	Switched-Capacitor Resistance
SDR	Software-Defined Radio
SNR	Signal-to-Noise Ratio
SPI	Serial Peripheral Interface
SoC	System-on-Chip
TDMA	Time Division Multiple Access
TX	Transmitter
UMTS	Universal Mobile Telecommunications System
USB	Universal Serial Bus
WCDMA	Wideband Code-Division Multiple Access
WLAN	Wireless Local-Area Network
ZOH	Zero-Order Hold
Z	Electrical Impedance

Biography

Pedro Emiliano Paro Filho was born in Araçatuba, Brazil, in 1984. He received the B.Sc. and M.Sc. degrees in Electrical Engineering from the University of Campinas, Brazil, in 2009 and 2012, respectively.

Pedro Paro joined IDEA Electronic Systems as a design engineer in 2009, working on (de)modulators for the ISDB-T standard. In 2010, he was an intern at the Interuniversity Micro-Electronic Centre (IMEC), Belgium, investigating sustaining amplifier topologies for MEMS-based oscillators.

In 2016, Pedro obtained his PhD degree with greatest distinction from IMEC and the Vrije Universiteit Brussel (VUB), for his research on charge-based transmitter architectures. His most significant research contribution includes the validation and implementation of an innovative digital transmitter architecture, with unique noise filtering capabilities. His interests involve different aspects of RF transceiver's design, including reconfigurable front-ends and low power radios.

Pedro Paro received the ISSCC 2015 Jan Van Vessem Award for Outstanding European Paper, granted by the IEEE Solid-State Circuits Society, for the work entitled "A transmitter with 10b 128MS/S incremental-charge-based DAC achieving -155dBc/Hz out-of-band noise."

Chapter 1
Introduction

1.1 The Fear of Disconnection

Communication is a central aspect in our lives, and it has always been. From the most simple nod to acknowledge a passing by colleague to a transatlantic file transfer between servers, we are always communicating, either verbally or non-verbally. However, during the last few decades society has observed an enormous shift on the way we communicate, work and socialize. Even the way we experience the world is being changed with an increasing adoption of "virtual" reality [Per16], to a point where the "digital" aspect in one's life became so crucial that the fear of being disconnected has gained its own denomination called *"Nomophobia"* [Kin13]. Despite the great deal of traffic created by less important content shared across social networks and other entertainment services, technology has been a great enabler to connect more and more people across the globe, and to distribute information in a more democratic way. Pretty similar to the Moore's law [Moo65], the famous human "Knowledge-doubling curve" theorized by Buckminster Fuller [Ful81] would simply not be possible if it was not for the facilitated means for people to communicate and exchange knowledge and ideas.

This communication network that is so widely spread among us relies on a huge infrastructure that has been deployed over the last century in order to enable fast and reliable exchange of data. Most of its communication links are "wired", but a great part of it is done over the air with no physical connection between the transmitting and receiving parts. From the early studies carried by Guglielmo Marconi, Reginald Fessenden and Lee de Forest,[1] wireless communication systems have evolved fast and tremendously. Satellite, backhaul and radio broadcasting are just few of the

[1]For an interesting read, the first chapter of [Lee03] offers a nice overview of the early days of wireless communications systems.

P.E. Paro Filho et al., *Charge-based CMOS Digital RF Transmitters*, Analog Circuits and Signal Processing, DOI 10.1007/978-3-319-45787-1_1

so many services occupying the frequency spectrum these days, and the push for ubiquitous connectivity has been so strong that even high speed Internet is now brought down to mobile handheld devices.

In fact, first designed to provide voice communication only, these mobile devices—now called "smartphones"—are so powerful today that it packs more computing power than a full featured server CPU sold less than 10 years ago. The number of wireless standards integrated has also increased significantly in order to support a wide range of connectivity and positioning applications, including GPS, WiFi, Bluetooth, NFC and etc. To our delight, having so many features in a battery-powered device with the additional constraints of cost, performance and size is a tough challenge in itself, fueled in this case not only by an increasingly connected population, but also a multi-billion industry that relies on continuous market growth to keep its gears turning.

1.2 Advanced Wireless Communication Systems

As indicated by Shannon's Law [Sha48], for a fixed Signal-to-Noise ratio (SNR) the channel capacity can only be increased by also increasing the transmit signal bandwidth. Since the early days of telecommunications, this has been a common ground to both wired and wireless systems as a way to increase communication speed. If we look at the cellular communication system as an example, the story has been pretty much the same.

Starting back in 1983, the first cellular communication systems (1G) were deployed around the world at a time where voice was the only user information being conveyed. AMPS, NTT, NMT were all analog-modulation-based wireless systems using mostly frequency modulation (FM) and duplexing (FDMA) to accommodate a very limited number of users [Cha01]. Almost unconceivable for today's standards, the channel bandwidth was only 30 kHz (AMPS), and a shocking 15 W user terminal transmit power was required to enable a cell radius of up to 40 km (NMT450) [Har06].

In the second generation, the era of digital cellular communication was inaugurated. Global System for Mobile communication (GSM) is still nowadays by far the most widely spread 2G technology, which applies time multiplexing (TDMA) to increase the user capacity and Gaussian Minimum Shift Keying (GMSK) as digital modulation. With a channel bandwidth of 200 kHz, a maximum 9.6 kbps data-rate can be achieved in its first version. Within the second generation still, what followed was a continuous upgrade of the GSM standard aiming at higher data rates while keeping the same channel bandwidth, first by increasing the number of time slots (in the GPRS), and later by improving the spectral efficiency with 8-PSK modulation (EDGE). Named 2.5G, the EDGE standard could provide a maximum data rate of 384 kbps.

Even though text messaging was already possible in 2G, it was only with the third generation (3G) that cellular communication changed from a pure telephony system

Table 1.1 Modern wireless communication systems

Modern wireless communication systems						
Standard		GPRS	EDGE	HSDPA	WLAN	LTE
Modulation		GMSK	8-PSK	QAM	64QAM OFDM	64QAM OFDM
Data rate[a]	[Mbps]	0.158	0.384	2	54	100
Channel bandwidth	[MHz]	0.2	0.2	5	20	20
Spectral[b] efficiency	[bps/Hz]	0.2	0.6	1.2	3.2	5

[a]Exact values may differ across literature
[b]Defined for BW60—occupied bandwidth 60 dB below peak [McC10]

to a packet-based network that could provide fast Internet access to mobile devices. Called Universal Mobile Telecommunication System (UMTS), the third generation applied wideband CDMA (WCDMA) as air interface, providing a maximum data rate of 2 Mbps with a bandwidth occupation of 3.84 MHz and channel spacing of 5 MHz. Also updated over time, in the latest 3G version—called High-Speed Downlink Packet Access (HSDPA), peak data rates of more than 10 Mbps can be achieved in theory.

The fourth generation (4G) moves into an all IP packet-based network, aiming to provide peak data rates up to 1 Gbps and 100 Mbps for high and low mobility access, respectively. The spectral efficiency is increased by using orthogonal frequency-division multiplexing (OFDM) and supporting multiple-input multiple-output (MIMO) access [GL10]. The channel bandwidth is increased to 20 MHz, but it can be further extended through carrier aggregation (CA). The idea is to overcome spectrum scarcity by combining multiple slots of 20 MHz inside the same or even across different bands, representing a significant design challenge these days.

Table 1.1 summarizes the key aspects of each standard, clearly showing the continuous increase in bandwidth and spectral efficiency. Wireless LAN is also included for comparison.

What will come after 4G is still pretty much undefined. However, it is widely agreed that the fifth generation (5G) should provide a 1000-fold increase in system capacity, as well as a tenfold improvement in spectral efficiency, energy efficiency and data rate (meaning incredible 10 Gbps and 1 Gbps for low/high mobility) [And14]. The aim is to achieve seamless and ubiquitous communication between anybody and anything (people to people, people to machine and machine to machine).

Due to spectrum scarcity, many applications will be pushed to higher frequency bands, where large and contiguous portions of the frequency spectrum are still available. Enabled by highly-scaled CMOS technology, an increasing amount of transceivers operating at the unlicensed 60 GHz as well as the E-band have been developed. For instance, two CMOS 60 GHz beamforming transceivers fulfilling the IEEE802.11ad requirements were successfully demonstrated in [Vid13] and

[Man16]. Channel bonding is also possible, as shown in [Oka14], allowing a data rate of 10.56 Gbps with 64QAM. Though evolving fast, at these frequencies however the challenges remain to provide the minimum required performance per block using CMOS technology (e.g. Phase Noise for 64-QAM), and implementing spatial power combining to overcome the increased atmosphere absorption.

1.3 Flexible Multi-Standard Operation

As for the lower side of the spectrum, circuit implementation is way more relaxed so the attention can be shifted to improving (mostly noise and linearity) performance and power efficiency. The continuous improvement of the radio frontend not only enabled small (battery-powered) handheld devices to keep up with the increasing complexity of wireless communication systems, but also becomes a necessity if looked from the system perspective, given the spectrum scarcity and the environmental impact caused by power-hungry base stations [Aue11].

From the user point-of-view, a lot of attention has been given to the design of transceiver architectures that could be used with multiple standards. Many devices such as smartphones and tablets provide connectivity to a large set of wireless standards (GSM, UMTS (3G), LTE (4G), Bluetooth, WLAN, GPS, etc.) and so far the typical implementation involves the simple hardware multiplication, impacting both size and cost, and providing no flexibility. To address this problem the concept of Software-Defined Radio (SDR) was created [Mit95], aiming at a highly flexible platform that could transform itself in order to satisfy the requirements of any communication protocol.

In this type of architecture, the multiple standard-specific hardware would be exchanged for a single radio frontend, whose performance would be reconfigured (by software) to support the reception/transmission of multiple standards, one at a time. High-end applications providing more speed or connectivity can also be foreseen, by just integrating more instantiations of the same frontend, all of them controlled with a single baseband engine [Glo03, Der09]. However, different standards impose completely different specifications on a device. For instance, GSM require large transmit power capability and stringent phase noise performance while bandwidth and data rate are very much relaxed. WLAN, on the hand, have lower transmit power requirements (since the modems are typically located within tens of meters), but bandwidth and data rates are increased in order to provide fast communication access. Therefore, a completely software-defined radio architecture would necessarily require a single TX and RX implementation that is capable of attending the most stringent requirements of each and every supported standard. Over the years, several transmitter [Par09, Yin13] and receiver [Bag06, Gia09, Bor13] implementations targeting SDR application have been demonstrated, including solutions working at both low and high cellular bands [Abi07, Cra07, Ing10].

For full flexibility however, and also compatibility with LTE carrier aggregation (CA), it is also desirable that the given radio frontend can operate across different bands. On the transmit side, it translates into even more strict TX noise specifications since typical bandpass (surface acoustic wave—SAW) filters typically applied to reduce the TX noise emission does not provide the required flexibility and hence cannot be used. Therefore, the design of single (and thus high performance) SAW-less transceiver architecture that can leverage power efficiency while being compliant with multiple standards still remains a hot research topic.

Technology is also an important aspect. In the early days, CMOS technology was not appropriate for radio frequency implementations, mostly due to a small breakdown voltage and reduced unity-gain current frequency (f_T). However, with the fast development of the semiconductor technology, not only the switching speed of CMOS has increased significantly, but the cost benefits provided with CMOS mass production has made it the perfect platform for high-end user products, including RF circuits. Moreover, for a cost effective solution it is also desired that the radio frontend can be integrated monolithically with the typically dominant digital processor, as in a System-on-Chip (SoC).

1.3.1 TX Frontend Key Requirements

Throughout the evolution of wireless communication systems, a clear migration has been observed from simple analog amplitude and frequency modulation to more complex and spectral efficient modulation schemes. First, analog modulation was exchanged by simple digital phase modulation (e.g. GMSK, OQPSK). Second, as from the third generation amplitude modulation was also introduced in order to increase the number of bits per symbol and thus improve the spectral efficiency. However, since the signal envelope is not constant anymore, non-linear distortion becomes a very important aspect.

Every communication standard has a limit on how much signal distortion can be withstood without affecting its performance. Typically, the linearity requirements are defined based on two aspects: out-of-band (OOB) emission and in-band signal integrity.

Ideally, transmitters should not generate or emit power outside its allocated frequency band. However, non-idealities of the several blocks comprising the signal path contribute to create different intermodulation products that end up outside the transmit band. These spurious emissions are seen as interference to the adjacent channels, and degrade their link quality. Thus, one way of assessing the TX linearity is by concurrently measuring the amount of power transmitted both in-band and at the adjacent channels (Fig. 1.1). The ratio between these two quantities defines the ACLR [Raz12]. Second, every communication standard defines a spectral mask (Fig. 1.1) that must be satisfied by the transmit signal, as another way to guarantee that interference is sufficiently small. In most cases, an increased out-of-band power emission or violated spectral mask is produced by excessive third and fifth-order distortion.

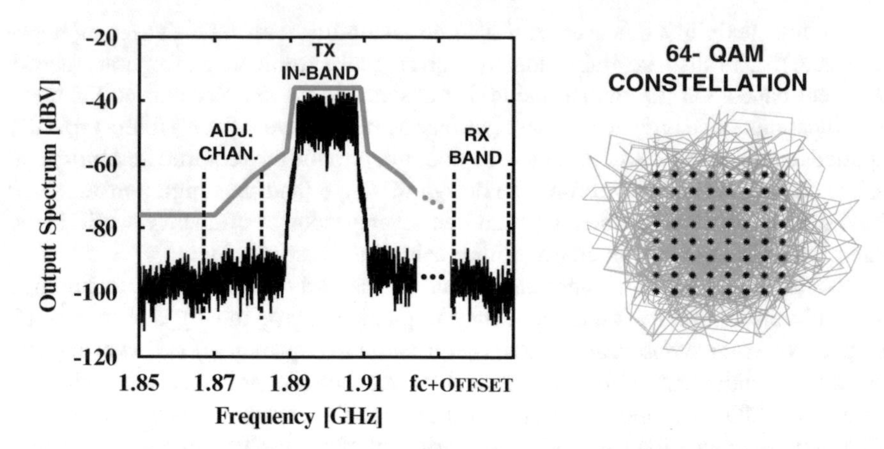

Fig. 1.1 Example LTE signal (*left*) showing spectral mask, adjacent channel and RX band (FDD), typically separated from the carrier (*fc*) by tens of megahertz (OFFSET). Captured modulated signal (*right*) with a EVM of 1.6 %

In-band signal integrity is another important performance parameter affected by distortion. In modern wireless systems where digital modulation is applied, the input data is translated into a limited set of symbols scattered around a constellation diagram. Each one of these constellation points correspond to a particular vector, whose distortion hampers the subsequent signal decoding and leads to an increased bit error rate (BER). The Error Vector Magnitude (EVM) is a common figure of merit used in this case to assess the amount of degradation caused by in-band signal distortion, among other contributors. Figure 1.1 shows a measured vector diagram with an EVM of 1.6 %. When the constellation density is larger (as in high-order modulation schemes), the minimum EVM required to achieve a sufficiently low BER is reduced, and hence more challenging.

Besides linearity, the introduction of non-constant envelope, high-order modulation schemes also impact the system power efficiency. It is known that large bandwidth modulated signals using high-order QAM modulation exhibit large envelope variations (Fig. 1.2), which translates into also large peak-to-average power ratios (PAPR) [McC10]. Simple modulation schemes generating constant envelope output signals (PAPR equal to 0 dB) allows the power efficiency to be increased since most blocks composing the signal path can be operated close to (or even above) saturation, maximizing power efficiency. Unfortunately, this is not possible with advanced wireless standards. To avoid compression of the signal peaks, the large PAPR commonly seen in high-order modulation schemes (Table 1.2) forces the transmitter to operate from 5 to 10 dB below saturation power (backoff), where the power efficiency is significantly reduced. A clear trade-off is observed in this case, reason why increasing efficiency without degrading the TX linearity is not straightforward. Among power amplifiers, Envelope Tracking (EP) [Oni13], Envelope Elimination and Restoration (EER) [Yoo12, Ois14] and Doherty topologies [Kay15] seem to be preferred methods to relax this trade-off in current literature.

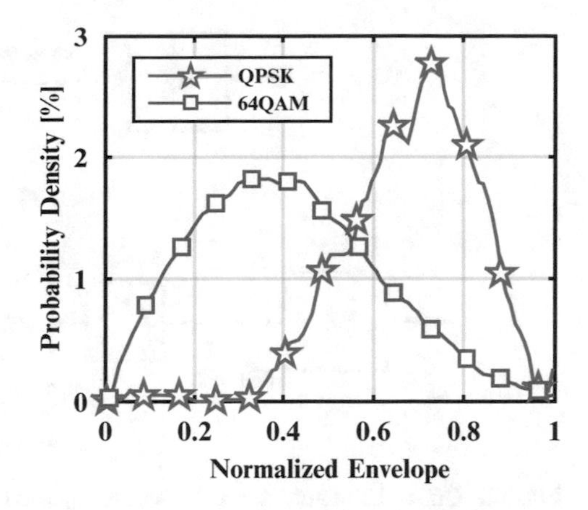

Fig. 1.2 Probability Density for QPSK and 64QAM modulations (*left*). The increased probability at small amplitudes also translates into large PAPRs, as shown in Table 1.2 (*right*)

Table 1.2 PAPR versus wireless standard

Standard	PAPR [dB]
GSM	0
EDGE	3.2
WCDMA	3.5–5
LTE	11
WLAN	12–16

When designing multi-standard transmitters for wireless communication systems, noise is obviously another important concern. Besides the aforementioned harmonic distortion, actual transmitters also emit out-of-band noise. In the case of GSM, the out-of-band (OOB) noise at the receive band should always remain below −129 dBm/Hz [Raz12], so that a transmitting mobile station does not interfere with a receiving one in close proximity. Considering the peak output power specification of 33 dBm, this requirement translates into a relative spectral noise density of −162 dBc/Hz, which is one of the most difficult specifications to be met in a GSM-compliant transmitter design.

Besides the interference between different users, the out-of-band noise can also be a problem within the same transceiver when frequency-division duplex (FDD) is used, in which case both TX and RX are active at the same time. As shown in Fig. 1.3, in this type of architecture transmitter and receiver are both connected to the antenna through a duplexer. Since the isolation between TX and RX is not infinite, a fraction of the transmit noise reaches the RX input, which is typically filtered with an interstage SAW filter between the modulator output and the PA. However, if SAW-less operation is targeted, very tough requirements are put on the intrinsic out-of-band noise performance of the transmitter. As shown in [Oka11], given a maximum acceptable noise power density of −178 dBm/Hz at the RX input, a duplexer isolation of 50 dB leads to a maximum PA output noise power of −128 dBm/Hz. With a PA gain of 27 dB and a duplexer insertion loss of −3 dB, to support an RMS output power of 24 dBm at the antenna the required carrier-to-noise ratio (CNR) at the PA driver output should remain below −155 dBc/Hz. As noted,

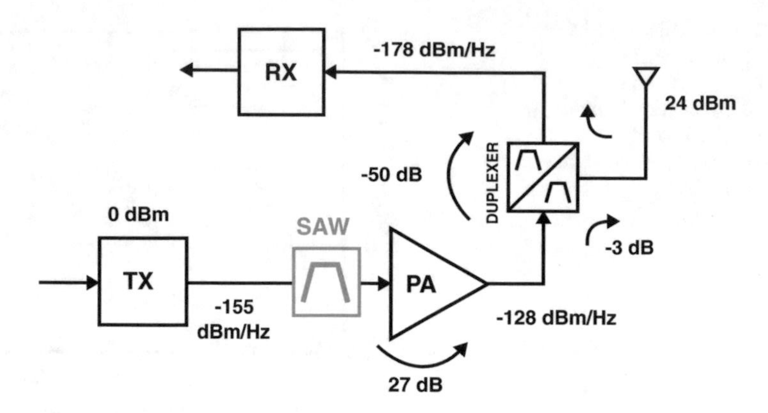

Fig. 1.3 Out-of-band noise calculations. Example extracted from [Oka11]

the actual CNR specification necessarily depends on the particular components being used, often leading to even more stringent out-of-band noise requirements.

In summary, with respect to the transmitter front-end, providing an extremely low noise emission with an improved TX linearity are key requirements to supporting advanced wireless communication standards without the aid of a SAW filter. Unfortunately, increasing power and/or area consumption is not an option in this case. On the contrary, for an improved user experience the battery should last longer, and the price per component should always go down, so that more and more features can be added to this mobile handheld device.

1.4 High Performance TX Architectures

Until very recently, roughly all transmitter implementations for wireless communications were analog intensive, typically based on a quadrature architecture, as shown in Fig. 1.4. The In-phase (I) and Quadrature (Q) components are the simple Cartesian representation of the transmit signal, which are first converted to analog domain using a digital-to-analog converter (DAC) with a typical zero-order hold (ZOH) signal reconstruction. Quantization noise and sampling aliases are then filtered with a low-pass reconstruction filter, before the baseband signal is upconverted into RF frequencies using a quadrature mixer. In typical cases, the upconverted RF signal is first amplified with a pre-power amplifier (PPA) before being fed to the PA, which is responsible for delivering the required output power to the antenna as defined by the wireless standard.

Historically, this architecture has always been preferred for integrated CMOS transmitters since it successfully delivered the required performance. However, it also comprises power-hungry and bulky components that, together with the poor analog characteristics of nanoscale CMOS technology nodes, served as a good

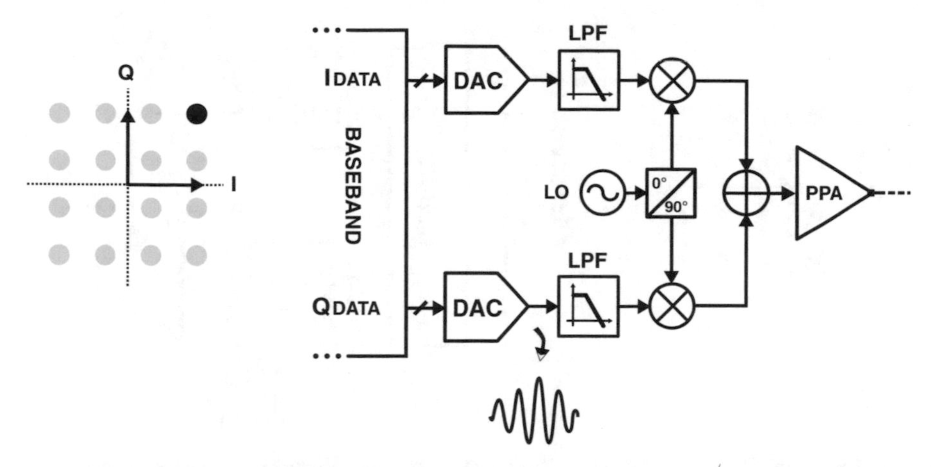

Fig. 1.4 Typical analog-intensive direct conversion quadrature TX

motivation for the introduction of more power and area efficient, as well as more digital and scaling-friendly architectures.

A widely accepted alternative to achieve better power efficiency is to use polar modulation, as seen in [Sta04, Sta05, Meh10, Lia13, You11]. In this type of architecture (Fig. 1.5), rather than using a quadrature I/Q representation, the transmit signal is decomposed into amplitude (A) and phase (ϕ) components [Eq. (1.1)], which can be processed and amplified separately and then recombined at the output [Gro07].

$$\begin{cases} A(t) = \sqrt{I^2(t) + Q^2(t)} \\ \phi(t) = \tan^{-1}(Q(t)/I(t)) \end{cases} \tag{1.1}$$

The combining operation is commonly performed by using the amplitude signal to modulate the output stage supply voltage, or controlling the number of ON current cells as in a digital PA (DPA) [Sta05, Meh10, Lia13]. The phase modulation, in turn, is typically done with a phase-locked loop (PLL). The main advantage of using polar modulation regards the great improvement in power efficiency that can be achieved. First, the constant-envelope phase information allows the entire phase path to be operated in saturation. Second, by modulating the supply voltage of the output stage, the architecture can provide improved power efficiency even at large backoff from saturation.

However, polar modulators also have important disadvantages when compared to the quadrature counterparts. First, from the non-linear derivation given by Eq. (1.1), the bandwidth of both amplitude and phase signals are largely expanded, requiring a similar increase in the corresponding circuitry's bandwidth and sampling frequency (4–6 times, according to [Ye13a]). Even though modern CMOS technologies can easily accommodate large switching speeds, this aspect can be problematic when the signal bandwidth is increased. Second, the architecture is inherently

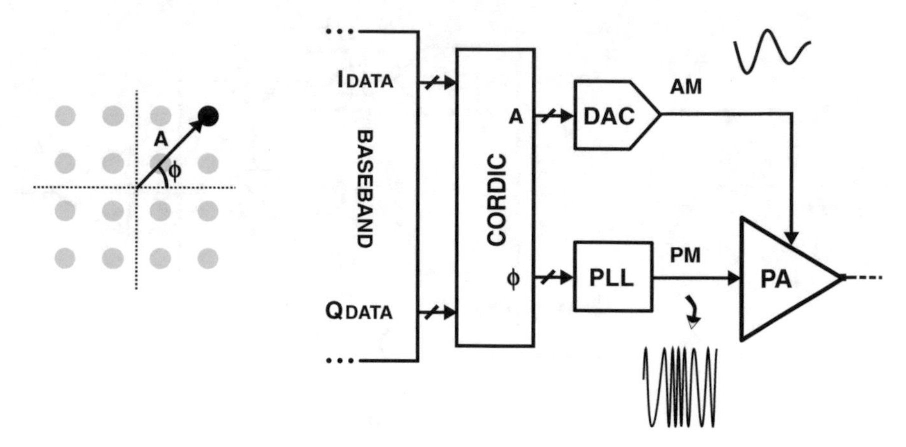

Fig. 1.5 Simplified polar architecture. The Cartesian to polar conversion is typically implemented using a coordinate rotation for digital computer (CORDIC) algorithm [Meh09]

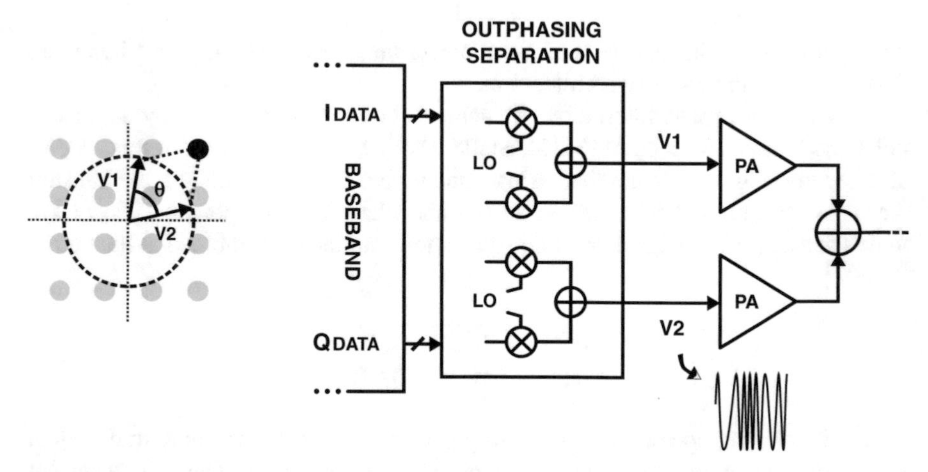

Fig. 1.6 Simplified Outphasing block diagram. Both V1 and V2 are constant-envelope signals that can be amplified using very non-linear PAs

asymmetric, and both amplitude and phase signals can have different propagation times. Either caused by delay mismatch or limited bandwidth, imperfections of the polar architecture can easily create unwanted expansion of the output spectrum, called spectrum regrowth. With some exceptions [Kav08, Cho11, Ye13b], these aspects are the main reasons preventing a wider application of polar transmitters in advanced wireless communication systems.

Another way to avoid the power efficiency degradation created by envelope variations is to decompose the transmit signal into two constant-envelope waveforms. Called outphasing [Chi35], in this type of modulation the wanted signal is expressed as the sum of two phase-modulated components (V1 and V2), outphased by θ, as shown in Fig. 1.6.

Similar to the polar implementation, the outphasing architecture can operate with completely nonlinear amplifiers, with the benefit that no supply modulation is required, and both signal paths are identical—and thus better matched. Besides the also increased bandwidth of both V1 and V2, one of the most relevant issues regarding the outphasing architecture is the fact the power combiner should combine the power of two amplifiers with different output signals. The output impedance variation at both amplifier outputs creates a time-varying voltage division and hence unwanted cross-dependence between both amplifiers, typically implying distortion [Raz12].

From the aforementioned analysis, it is concluded that both polar and outphasing architectures can be very well suited as alternatives to increase the power efficiency of both transmitter and PA at large backoff conditions. However, it is also true that achieving outstanding linearity using these architectures is quite challenging. For this reason, the work presented in this book utilizes a quadrature direct-conversion TX architecture, looking into ways of improving its noise performance and power efficiency, while providing the required linearity.

1.5 Quadrature Direct-Conversion Transmitters

The literature analysis shows that with regard to quadrature transmitter implementations, the state-of-the-art is clearly divided into analog and digital-intensive architectures. During the last decade, a clear trend was observed where an increasing number of digital-intensive transmitters have been developed, as a way to survive and perhaps benefit (in few aspects) in a digital-driven technology environment.

Also because of its historical prevalence, analog-intensive architectures provided so far the best performing implementations. However, their improved noise performance is typically achieved through extensive low-pass filtering along the entire signal path. Upconverted baseband noise has a very significant impact in the out-of-band noise, reason why large and bulky reconstruction filters are typically placed after DAC conversion, not only to reduce the sampling aliases, but also to filter out the multiple noise contributors, including quantization noise (Fig. 1.7).

In [Gia11], large flexibility is achieved using a Tow-Thomas filter implementation that offers independent programming of transimpedance gain, bandwidth and quality factor. In [Cas09], an active inductor providing a second-order low-pass bi-quadratic transfer function is used to improve noise filtering, not without an impact in power consumption. On top of reconstruction filtering, the work of [Mir08] applies a feedback filtering technique that introduces a null with an arbitrary width at the receive frequency, achieving -160 dBc/Hz at 80 MHz offset. Common to all the aforementioned examples is the considerable area consumption contributed by the reconstruction filters when SAW-less operation is targeted. For instance, in [Oli12] 1.37 mm^2 is taken by the baseband filter alone, in a 90 nm technology. Besides the area consumption, the introduction of a reconstruction filter in the signal path also

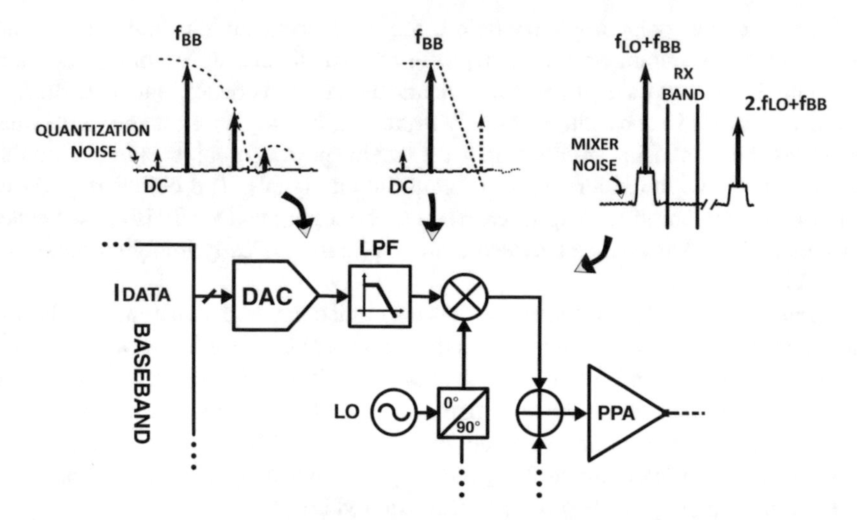

Fig. 1.7 Spectrum content at the different stages of conventional analog-intensive I/Q transmitters. The reconstruction filter significantly attenuates both noise and spurs at the RX-band

increases the path loss toward the mixer input, and reduces the effectiveness of pre-distortion by filtering high frequency components added on purpose to compensate for frontend non-linearities.

The choice of mixer topology has also an important impact in out-of-band noise performance. A large number of analog-intensive implementations apply traditional current mode Gilbert-cell active mixers to perform baseband upconversion. However, the noise contribution of the several transistors composing the mixer topology cannot be easily filtered in current domain without using bulky (and thus expensive) inductors. A SAW-less transmitter using a Gilbert-cell mixer would hence require an intrinsically low noise mixer design, which would finally result in a very large power consumption, with an output current comparable to what is delivered by the pre-power amplifier. Therefore, in a current-mode design a good practice is to combine the mixer and the pre-power amplifier into one single block to maintain power efficiency [Jon07, Cod14]. In general, operating in voltage domain and using a voltage sampling mixer and a pre-power amplifier [He09, Col14] can be more power efficient, since the noise contribution of the mixer switches can be easily made negligible.

Another problem faced by analog architectures in general is the reduced portability and constant deterioration of the analog characteristics observed in highly-scaled digital-driven CMOS technologies. The ever decreasing supply voltage and intrinsic gain, combined with increased leakage (or alternatively large threshold voltage) make the design of high performance analog-intensive topologies more difficult in every new technology node.

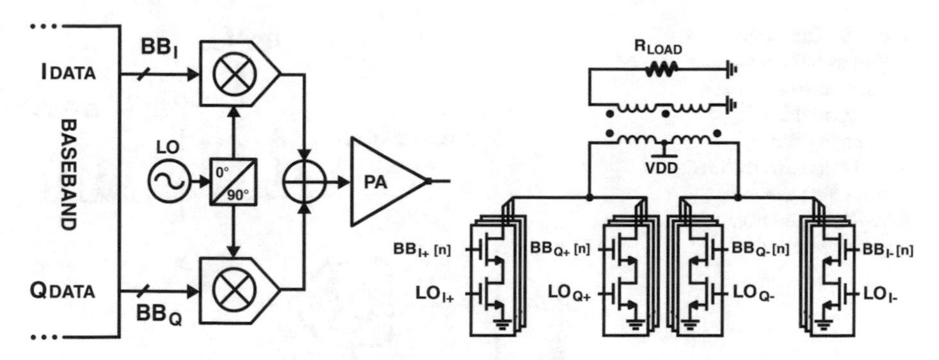

Fig. 1.8 Simplified direct-conversion digital-intensive I/Q transmitter (*left*). Current-based DACs combined with the LO signal provide direct digital-to-RF conversion (*right*) as proposed in [Elo07]

As the proverb says: "If you cannot beat them, join them", and by applying this logic a new class of architectures is created, where the analog share of the TX frontend is intentionally reduced to a minimum. Named digital-intensive transmitters, these architectures are undoubtedly more portable, area efficient, and robust to the aforementioned analog impairments of the technology. The benefit of compact implementation can be clearly noted in almost every digital-intensive implementation. For example [Lu13] is able to achieve a maximum output power of 24.7 dBm consuming only 0.5 mm^2 in 40 nm technology. The works presented in [Ing14] and [Ala13], can also achieve the reduced area consumption of 0.32 mm^2 and 0.6 mm^2 respectively, including the integrated baluns.

In a typical quadrature direct-conversion digital-intensive transmitter (Fig. 1.8), the DAC and the mixer are combined into a single entity called in many cases RFDAC, which provides at once both functionalities of digital-to-analog conversion and mixer up-conversion.

The Direct Digital-to-RF Modulator (DDRM) proposed in [Elo07] and [Ing14], and the RFDAC implemented in [Ala12] are great examples of this type of architecture. They combine several current sources in a steering DAC-like structure, where the number of current sources switched together is locally controlled and determined by the digital code, while the LO switching provides the signal up-conversion. Since the digital baseband signal is directly brought to the mixer switches, the transmitter is more robust to imperfections and mismatches, like DC-offsets and I/Q imbalances.

Digital-intensive transmitters would simply be the ultimate solution if it was not for their typically increased noise and spurious emission. The direct digital-to-RF conversion leaves no analog path where the reconstruction filter is typically introduced. As a result, both quantization noise and sampling aliases are upconverted to RF frequencies many times without any attenuation (Fig. 1.9), making it more challenging for this type of architecture to meet the stringent spur and out-of-band noise requirements of SAW-less operation.

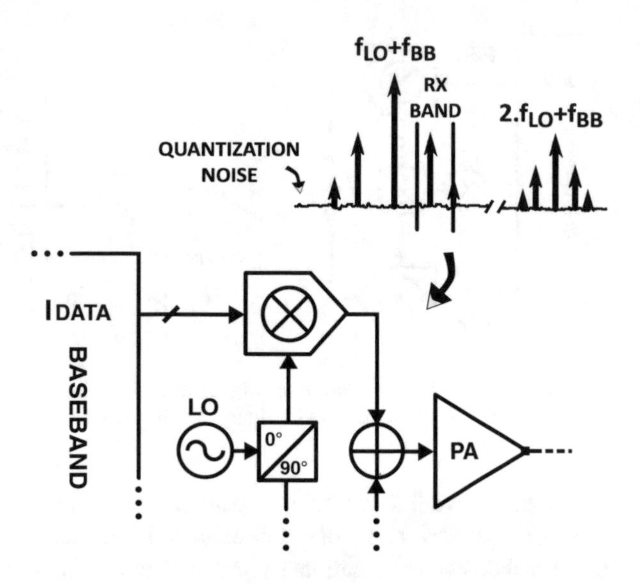

Fig. 1.9 The direct digital-to-RF conversion hinders noise and spurs filtering, making it challenging for digital-intensive architectures to meet the stringent SAW-less operation requirements

A common way of reducing the impact of quantization noise is by increasing the DAC number of bits and/or sampling frequency. In [Boo11] for instance, a 17-bit equivalent resolution is achieved with a 14-bit DAC using a GHz range clock. Other implementations apply $\Sigma\Delta$ modulation to shift the quantization noise to larger frequencies [Jer07, Poz08, Fra09], but this solution can be inappropriate when the interstage SAW filter is removed, specially in the case of coexistence with other connectivity and navigation wireless services. Finite impulse response filters (FIR) [Gab11, Fuk12] are also commonly seen among other solutions as a way to reduce the quantization noise contribution at specific parts of the spectrum (e.g. RX-band in FDD operation).

Sampling aliases are also hardly attenuated at RF frequencies without the aid of SAW filters. In typical cases, high oversampling ratios (OSR) are used to move the aliases to higher frequencies and thus exploit the filtering effect of the DAC sin(x)/x (sinc) response [Boo11, Jer07], increasing power consumption and/or system complexity. Once the sampling frequency is increased, different types of filtering [Har03] and interpolation [Yij03b, Yij03a] can be applied to reduce the impact of the aliases in the out-of-band emission. Nevertheless, depending on the overhead circuitry and speed, some of these solutions can be conflicting with the intended reduction in power and area consumption.

From the portability point-of-view, a design is made more flexible and scalable by increasing its digital content. However, from the performance perspective, it seems inevitable that some sort of analog filtering is also included to guarantee that noise and spurs are sufficiently filtered, without recurring to solutions that may incur larger power and/or area consumption.

In this book, an innovative digital-intensive transmitter architecture is presented, which can provide the best of both worlds. It offers an improved robustness to technology impairments, high portability and reduced area in combination with a sufficiently low spurious and noise emission, so as to support SAW-less RF transmitter implementations.

The remainder of this book is organized as follows: Chap. 2 provides an introduction to the innovative concept of incremental charge-based operation. The main reasoning behind the charge-based operation and its operating principle are first explained using a watermill mechanical analogy. Second, with a "black-box" implementation of the envisioned charge-based architecture, the main benefits as well as the most important aspects involving noise and linearity performances are discussed.

Chapter 3 discusses the design and measurement results of the first realization of an incremental charge-based transmitter, based on a capacitive charge-based DAC (CQDAC). In a top-down approach, first the operating principles are discussed with the derivation of the required charge calculations. Noise and linearity performances are analysed considering the specific implementation, with special attention to all the noise filtering and alias attenuation capabilities intrinsically provided by the architecture. A complete discussion of the most important design aspects showing key layout details is also provided. In the last section, measurement results are provided, showing the sensitive improvements in noise performance achieved with the charge-based architecture.

Following the same structure of the previous chapter, Chap. 4 introduces the second charge-based TX realization, implemented using a resistive QDAC. In this implementation, the ability of delivering more power using the charge-based architecture was investigated. Based on the observation that the first chip's power consumption was highly impacted by the PPA bias current, a direct-launch implementation where the PA is directly driven using the QDAC was targeted. Again, the most important design aspects affecting noise and linearity performances are disclosed, with a complete description of the design of the several block comprising the TX implementation. Measurement results are provided at the end of the chapter, followed by a comparison table situating the results achieved in the current state-of-the-art. Finally, Chap. 5 concludes this book with a summary of the architecture's key benefits and results achieved with the two charge-based prototypes.

Chapter 2
Incremental-Charge-Based Operation

2.1 Introduction

To ease the understanding of the underlying concept of charge-based operation, let's take a mechanical system as an example. Imagine a watermill, which has to be controlled in a precise and timely manner following a wanted input command (Fig. 2.1).

Controlling the rotating wheel and hence the amount of power transferred to the structure is done in this case by controlling the amount of water that flows through the water pipes.[1] A larger water flow implies more energy being transferred to the watermill, and vice-versa.

Such control mechanism could be implemented in many different ways. For instance, let's consider a voltage controlled water pump, connected to the structure using pipes and taking water directly from the water main, as depicted in Fig. 2.2.

To assure the wanted functionality, first this specially designed pump should be strong enough to deliver the required flow to drive the rotating wheel at full power. Second, it should be able to resolve significantly small flow levels, so that the amount of water transferred to the wheel can be precisely controlled, and third, the output water flow should be—desirably—a linear function of the input control parameter, meaning that if $V_{CONTROL}$ is doubled, the water flow should also be doubled. At last, when transitioning from one flow level to another, the water pump should respond timely and consistently.

As one may expect, all of this requirements have a direct impact on how well the watermill can be operated, and therefore how close the output power resembles the

[1]For the sake of simplicity, a linear relationship between output power and the water flow is considered.

P.E. Paro Filho et al., *Charge-based CMOS Digital RF Transmitters*, Analog Circuits and Signal Processing, DOI 10.1007/978-3-319-45787-1_2

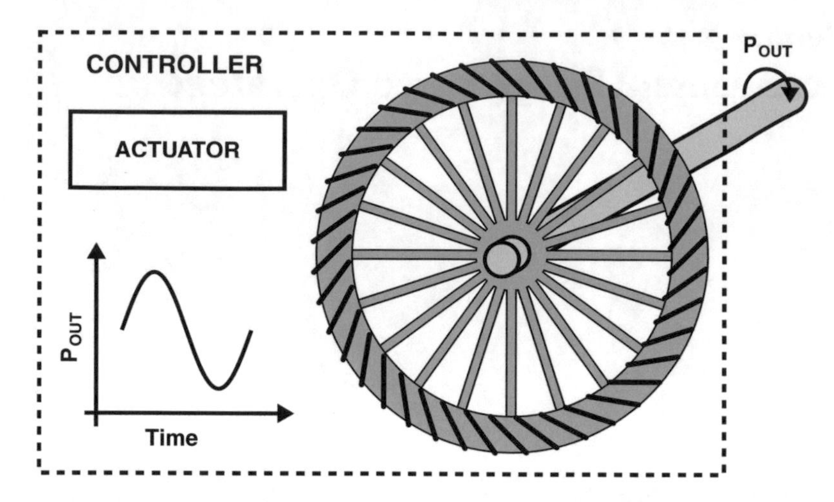

Fig. 2.1 Watermill mechanical system. The amount of power transferred to rotating wheel should follow a wanted input command

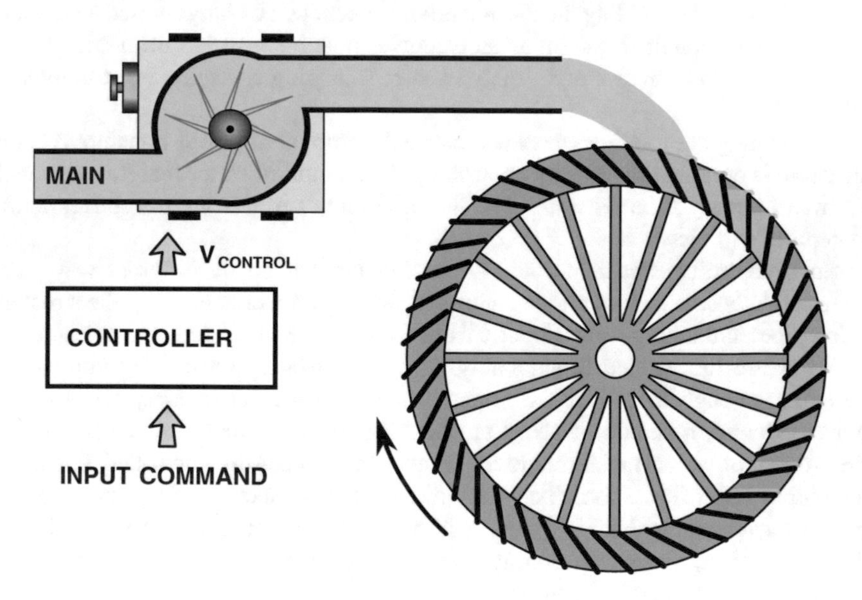

Fig. 2.2 Example control mechanism. A voltage-regulated water pump is used to control the amount of water pushed through the rotating wheel

wanted input command. Providing altogether a high power capability with a good degree of linearity and small enough intermediate steps is a tough challenge in itself, made even tougher when high power efficiency is required.

Since power consumption in typical cases can be dominated by fixed contributors that do not scale with the wanted output power, working at a fraction of the full capacity can lead to a considerable degradation of the system power efficiency

[Eq. (2.1)]. In our analogy, it means that most of the time the mechanical system will be running at poor efficiency levels, degraded by the large power consumption required to operate the strong water pump, compared to the reduced output power delivered.

$$Efficiency = \frac{P_{OUT}}{P_{PUMP}} = \frac{P_{OUT}}{\underbrace{P_{FIXED}}_{\text{DOMINANT}} + P_{VAR}} \quad (2.1)$$

Now let's assume the situation shown in Fig. 2.3. Instead of using a power hungry electrical pump to drive the watermill, the water pipes leading to the rotating wheel are now connected to a reservoir, whose water level can be directly manipulated with a featured controller device. Though the power transfer mechanism remains unaltered, now the control of the rotating wheel is implemented through the modification of the water level inside the reservoir. If the water level is increased, more water is pushed through the pipes—increasing the output power. Similarly, a decrease in water level incurs less power being transferred.

A top-level description of the system operation is depicted in Fig. 2.4. Based on the instantaneous input command $P_{OUT}[k]$, the water level ($W_{RES}[k]$) corresponding to the required flow is first calculated. The amount of water that should be added

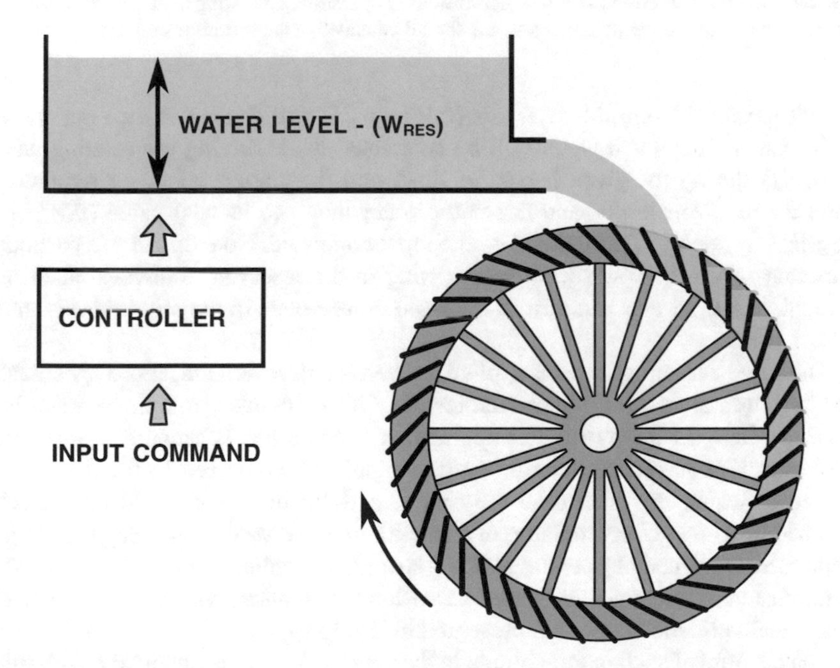

Fig. 2.3 Alternative control mechanism. Instead of using a water pump, the water flow is controlled by changing the water level in a large reservoir

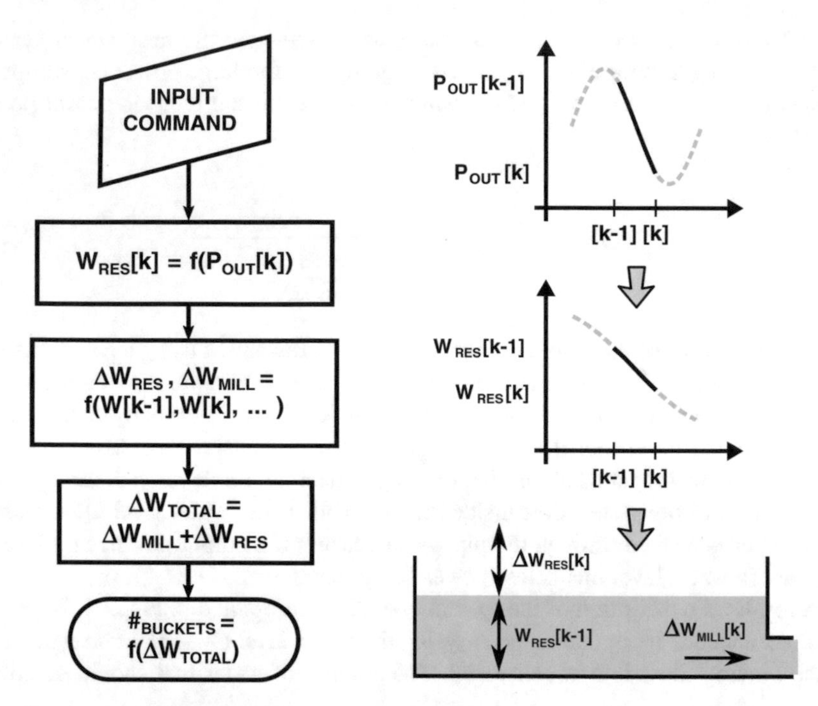

Fig. 2.4 Top-level description of the reservoir-based operation. Following the input command, the control system determines how much water should be added or subtracted from the reservoir

to (or subtracted from) the reservoir ($\Delta W_{TOTAL}[k]$) will depend on two parameters in this case: How much water will be subtracted while driving the rotating wheel (ΔW_{MILL}) during the given period of time, and the amount of water required to bring the reservoir level—and hence the water flow—to its next value ($\Delta W_{RES} = W_{RES}[k] - W_{RES}[k-1]$) as defined by the input command. Note that in this particular operation mode the water level pre-existing in the reservoir is always taken into account, and the water taken from the main corresponds to the required increment only.

The same reasoning can be applied in the electrical domain, more specifically into an alternative transmitter architecture. What in this example represents a rotating wheel, in the transmitter application it could be exchanged for an output (RF) load. The physical quantity being manipulated would be electrical charge—instead of water, in which case the role of a water reservoir would be perfectly matched by a charge accumulator, or capacitor in other words. Applying the already mentioned incremental operation, this transmitter architecture could be therefore denominated an incremental-charge-based implementation, whose exploration and conceptual validation were first presented in [Par15a].

The benefits of such modifications in the way the system is operated are clarified once the implementation of the featured water level controller is presented: As depicted in Fig. 2.5, imagine that instead of providing the reservoir with discrete

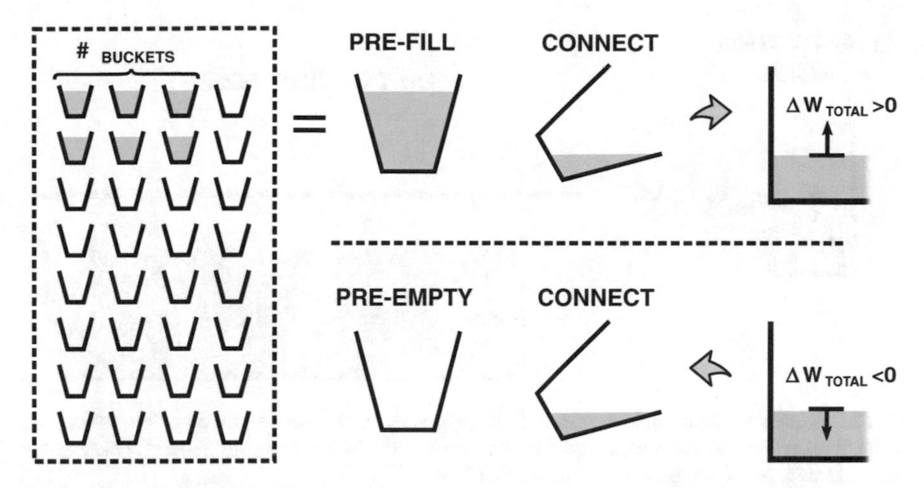

Fig. 2.5 Example structure used to control the reservoir level. According to required ΔW_{TOTAL}, the corresponding number of buckets ($\#_{BUCKETS}$) are pre-filled or pre-emptied before being connected to the reservoir, therefore increasing or decreasing the total amount stored

quantities of water using a water pump, a sufficiently large set of equally sized buckets are made available, which can be independently selected and pre-filled (or pre-emptied) before being connected to the actual reservoir.

If the water level is to be increased, the "configurable"-size bucket is first filled with water, which is later transferred to the reservoir. Similarly, when the level is to be decreased, the buckets are first emptied before connecting. Whenever a different input command is received, the corresponding bucket size is determined, as well as whether it should be pre-filled or pre-emptied, so that the required amount of water is provided. Notably, using buckets to change the water level in the reservoir can have a very interesting impact in the overall system operation and performance:

First, the minimum change in the water level that can be produced in the reservoir—and hence in the water flow—is determined by and scales with the smallest bucket available in the water manipulator. If the smallest bucket size is divided by 2, the minimum increase/decrease producible in the water level is also halved. In fact, the resulting change in water level is also inversely dependent on the reservoir size itself, meaning that for a fixed minimum bucket size, the increment in water level can also be halved by doubling the size of the reservoir. Therefore, rather than an absolute dependence on the minimum bucket size, the water flow resolution is proportional to the ratio between the smallest bucket and the reservoir size, as depicted in Fig. 2.6 and demonstrated in Eq. (2.2).

$$\Delta W_{MIN} \propto \frac{\text{Minimum Bucket Size}}{\text{Reservoir Size}} \Rightarrow \left| S^2_{nQUANT} \right| \tag{2.2}$$

Compared to the pump-based implementation, a ratio-dependent minimum step simplifies immensely on achieving the required water-flow resolution. In the

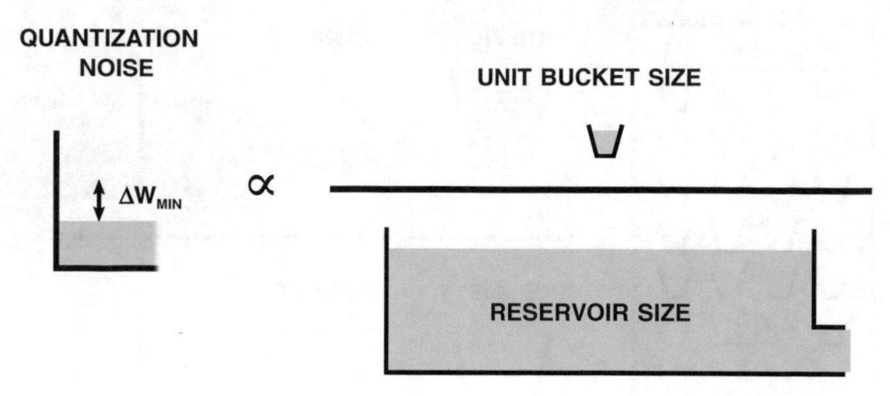

Fig. 2.6 Minimum achievable water level in(de)crease. In this alternative implementation, the water level resolution (and hence quantization noise) can be improved by either increasing the reservoir size, or reducing the minimum bucket size

transmitter domain, it represents the ability to reduce the quantization noise floor (S^2_{nQUANT}) below the stringent out-of-band noise requirements of advanced wireless systems by simply reducing the ratio between two capacitors. This feature is further explored in Sect. 3.2.3.2.

Second, feeding the watermill through a reservoir provides intrinsic filtering capabilities to the structure. Whilst in the pump-based solution variations of multiple sources (e.g. changes in the pump supply voltage, water main pressure, etc.) will have a direct impact on the output water flow, the application of a water reservoir provides an intrinsic buffering effect where all sorts of disturbances and fluctuations ("noise") are naturally averaged and therefore attenuated by the structure. In fact, the larger the reservoir the more damping is provided, further attenuating fluctuations in the observed water flow. In the context of transmitters for advanced wireless communication systems, this characteristic is a key enabler to achieve a sufficiently low out-of-band noise emission, filtering out from supply-coupled to quantization and thermal noise. A more complete explanation of the intrinsic noise capability of the charge-based structure is given in Sect. 2.2.1.

Third, different from typical architectures the charge-based operation can also provide utmost efficiency when operating at lower power levels. First of all, there is no "fixed" consumption, and water is only consumed when the reservoir level should be increased. Second, since the amount of water pre-existing in the reservoir is always taken into account, only the required increment (ΔW_{TOTAL}) is taken from the main. The same reasoning can be applied to the incremental charge-based TX architecture. The amount of charge taken from the supply—and hence the power consumption—scales with the signal amplitude, inherently improving power efficiency at backoff conditions.

Finally, one may argue that driving a large water reservoir (or charge accumulator) can also impact the overall system power efficiency. However, the benefits provided by the intrinsic noise filtering capability allows the reduction of both system complexity and power consumption related to signal noise filtering. In other words, in the charge-based operation the trade-off between noise performance and power consumption is significantly leveraged, as further clarified in the following sections.

2.2 Incremental-Charge-Based Operation

Suppose that instead of driving a watermill, now an electrical load Z is to be driven following a given input command (Fig. 2.7). Again, the power transfer from the supply to the output load has to be done in a precisely-controlled and timely manner, demanding sufficient linearity and bandwidth.

In a conventional voltage-output driving stage, more bandwidth is necessarily provided with a corresponding increase in transconductance. Circuit linearity, on the other hand, is typically leveraged by increasing the voltage headrooms and supply voltages whenever possible. In both cases, when stringent bandwidth and linearity requirements apply, the resulting DC power consumption implied by large bias currents and/or supply voltages can significantly impact the overall system efficiency.

Looking at the electrical nature of Z, a clear trend can be noticed though. In many cases, the output load to most blocks in any particular signal path is purely capacitive, given by layout parasitics and gate capacitances of the succeeding block's input transistors. As in the watermill example, these capacitances function as reservoirs whose charge content could be manipulated incrementally. Therefore, instead of driving the output load with a conventional voltage-output Digital-to-Analog Converter (DAC), the output signal in this architecture would be constructed by adding or subtracting charge from the output load as shown in Fig. 2.8, taking into account the previous amount of charge existing at all times.

The operating principles are very simple. Assuming for a first example a purely capacitive load (C_{LOAD}), it is possible to calculate how much charge should be moved to the output (per sampling period) based on the input command, as shown in

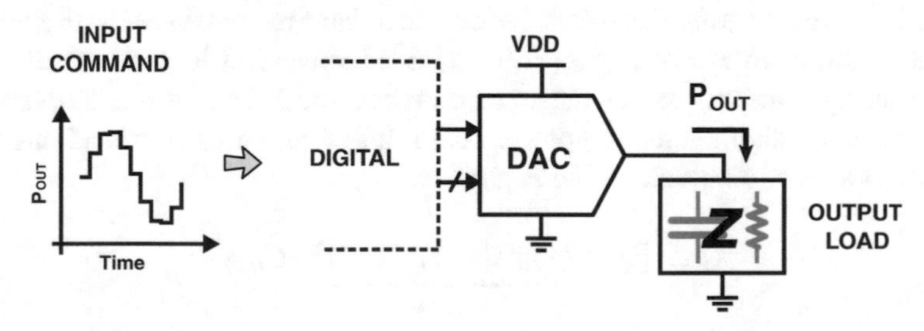

Fig. 2.7 Simplified representation of a typical system where a load Z has to be driven following a wanted (digital) input command

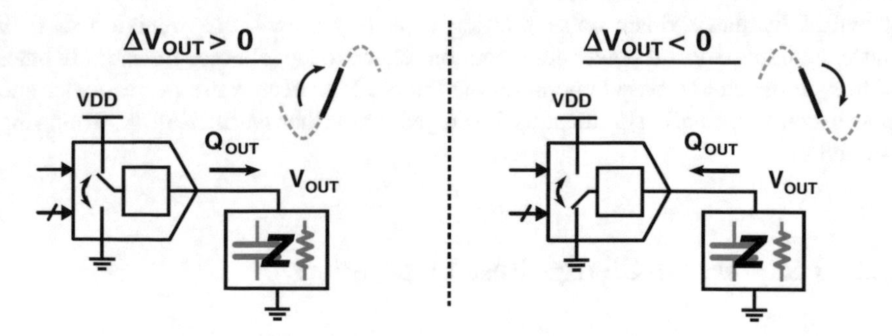

Fig. 2.8 In the incremental charge-based architecture the output voltage is changed by adding and subtracting charge from the output load

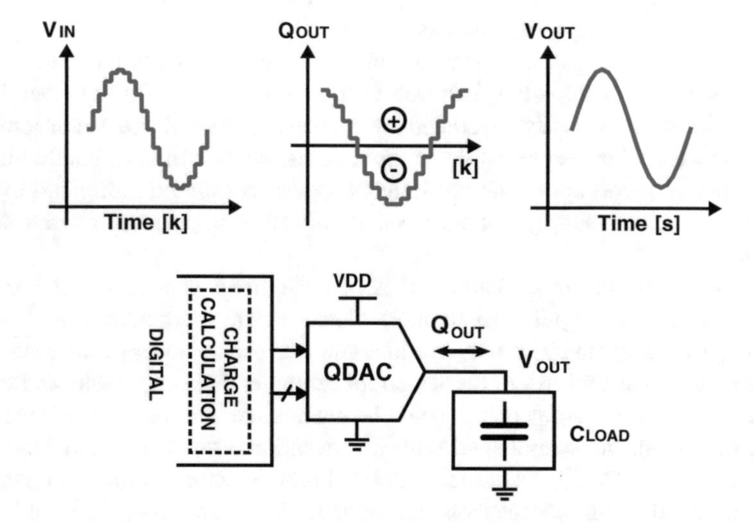

Fig. 2.9 Charge-based operation. A charge calculation block defines how much charge should be moved to/from the output load

Eq. (2.3). If the total amount of charge stored in C_{LOAD} is to be increased, charge is pulled from the supply, otherwise the excess charge is drained into the ground. These calculations are performed by a calculation block that first translates the digital input command into charge values (Fig. 2.9), and then converts it into the specific DAC input quantity (current, conductance, capacitance, etc.). Due to the simple nature of the required calculations, the power consumption and area overhead due to the additional calculation block can be negligible.

$$Q_{OUT}[k] = \underbrace{(V_{IN}[k] - V_{IN}[k-1])}_{\Delta V_{IN}} \cdot C_{LOAD} \tag{2.3}$$

The required charge Q_{OUT} can be delivered in different ways through what is called a charge-based DAC—or *QDAC*, either using discrete packets of charge conveyed to the output at fractions of the sampling period, or in continuous-time by constantly charging and discharging the output load throughout time.

Now imagine that this charge-based DAC is capable of providing the required bandwidth and linearity without requiring any biasing. Without a fixed bias current, the incremental operation assures that the power consumed to drive the output load is directly proportional to the output signal's amplitude. In other words, the DC power consumption of the charge-based architecture scales with the useful output power, leading to an improved power efficiency even at backoff conditions. In fact, for the simple example shown in Fig. 2.9, if the power spent with the QDAC operation (say due to digital switching) is made negligible, the charge-based DAC current consumption for a given output capacitance C can achieve the theoretical limit of fCV coulombs per second (or ampères)—where f and V are the output signal's frequency and amplitude, respectively.

The charge-based operation is not limited to pure capacitive loads, however. As discussed in Chap. 4, it is also possible to apply the same reasoning when driving a resistive output load and still observe all the benefits of charge-based operation, including improved efficiency and intrinsic noise filtering capabilities, as discussed below.

For the sake of a better understanding, the charge-based architecture and its intrinsic benefits are discussed here using a generic "black-box" charge-based DAC implementation. Further details about the two different implementations studied in this book are given in Chaps. 3 and 4.

2.2.1 Noise and Alias Performance

2.2.1.1 Alias Attenuation

In typical Digital-to-Analog Converters (DAC), the output voltage (or current) is a simple translation of the digital input code into a fraction of the converter's output full-scale. In a conventional implementation, the output signal remains fixed when the digital input word is not changing, corresponding to a zero-order hold (ZOH) [Opp97].

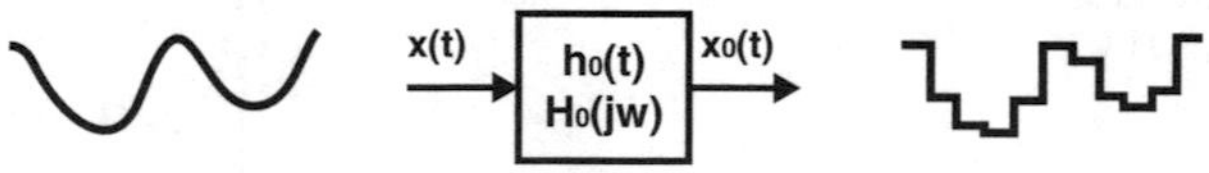

x(t) $h_0(t)$ $H_0(jw)$ $x_0(t)$

$$H_0(j\omega) = e^{-j\omega T/2}\left[\frac{2\sin(\omega T/2)}{\omega}\right]$$

$$= Te^{-j\omega T/2}\left[\frac{\sin(\pi\omega T/2\pi)}{\pi\omega T/2\pi}\right]$$

$$= Te^{-j\pi(\omega/\omega_0)}\,\mathrm{sinc}\left(\frac{\omega}{\omega_S}\right) \tag{2.4}$$

where $\omega_S = \dfrac{2\pi}{T}$ and $\mathrm{sinc(x)} = \dfrac{\sin(\pi x)}{\pi x}$.

As demonstrated in Eq. (2.4), the output spectrum of a converter applying zero-order hold is shaped by a sinc(x) function, notching at every multiple of the sampling frequency. The sampling aliases, centered around $n \cdot \omega_S$ (where $n = [1, 2, 3 \ldots]$), are also shaped by the same sinc(x) function, as shown in Fig. 2.10.

More involved converter architectures apply different interpolation schemes in order to further attenuate the sampling aliases. In most cases, either the input data is oversampled and interpolated using digital filters [Yon07, Yij03b], or multiple copies of the same converter are interleaved and operated at different phases of the sampling clock [Chi10]. While the former solution increases power consumption with digital operation at high clock frequencies, the latter can have a significant impact in area consumption.

Consider the case of a charge-based architecture where the charge transfer between supply and output load does not happen instantly, but rather uniformly distributed along the entire sampling period. If a constant amount of charge is delivered to a purely capacitive load, the output voltage increases linearly over time, corresponding in this case to an inherent first-order interpolation.

Different from a zero-order hold system, the application of linear interpolation has a beneficial impact on the reconstructed output spectrum. Demonstrated in Eq. (2.5), instead of a sinc(x), the aliases in this case are shaped by a $\mathrm{sinc(x)}^2$ transfer function, significantly reducing the sampling aliases as shown in Fig. 2.11.

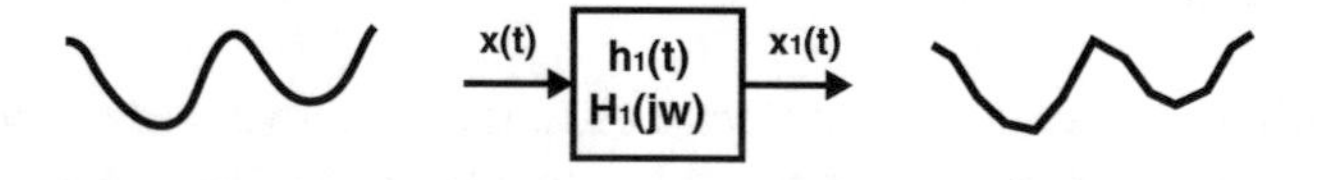

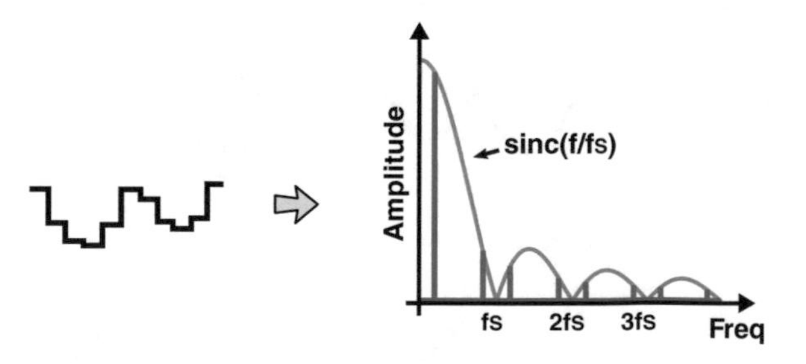

Fig. 2.10 Using zero-order hold the reconstruction spectrum is shaped by a sinc(x) function

Fig. 2.11 $\text{sinc}(x)^2$ alias attenuation

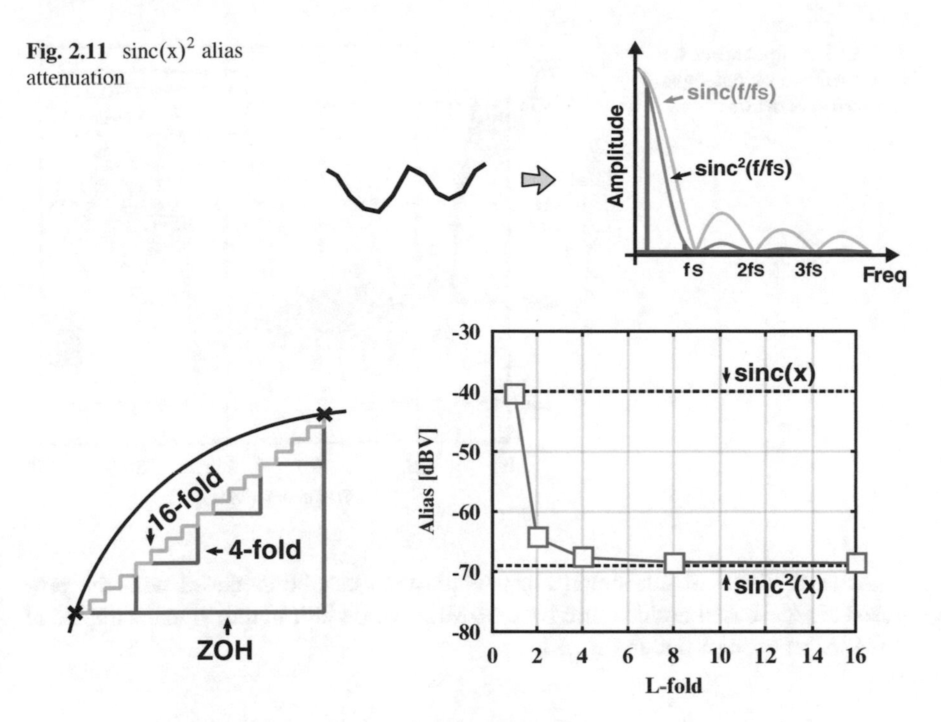

Fig. 2.12 L-fold linear interpolation for different values of L

$$
\begin{aligned}
H_1(j\omega) &= \frac{1}{T}\left[\frac{\sin(\omega T/2)}{\omega/2}\right]^2 \\
&= \frac{1}{T}\left[\frac{T\sin(\pi\omega T/2\pi)}{\pi\omega T/2\pi}\right]^2 \\
&= T\,\text{sinc}^2\left(\frac{\omega}{\omega_S}\right)
\end{aligned}
\tag{2.5}
$$

Even when the charge conveyed to the output node is not uniformly delivered over time, but rather in multiple sub-steps at a fraction of the sampling period (as in a L-fold linear interpolation [Yij03b]), yet the output spectrum shaping approaches the $\text{sinc}(x)^2$ transfer function for values of L above 2, as demonstrated in Fig. 2.12 for 20 MHz output single-tone sampled at 500 MS/s. It should be noted that transmitting charge at a higher speed in this case does not imply oversampling the input signal or interleaved operation. The calculation engine and the corresponding interface toward the DAC input are all driven at the same speed as the digital input data (F_S).

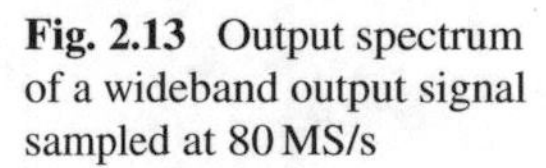
Fig. 2.13 Output spectrum of a wideband output signal sampled at 80 MS/s

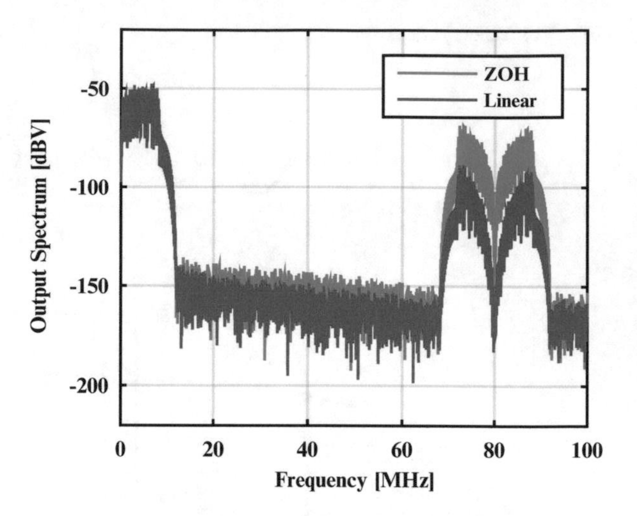

At least 30 dB of additional alias suppression can be achieved with the proposed charge-based architecture for a 20 MHz bandwidth output signal sampled at 80 MS/s, as exemplified in Fig. 2.13.

2.2.1.2 Intrinsic Noise Filtering

Besides the inevitable quantization noise, multiple noise sources also impact the output signal in a typical DAC implementation. Depending on the actual implementation, thermal and flicker noise from the internal DAC components, or even noise coupled to the supply and/or reference voltage can also contribute to a Signal-to-Noise Ratio (SNR) degradation. In cases where the noise spectral density at specific parts of the output spectrum should be kept at a minimum, or when stringent spectral masks should be met, typically a reconstruction filter is applied in order to attenuate both output noise and sampling aliases after digital-to-analog conversion (Fig. 2.14).

An important drawback of using reconstruction filters is the large area typically taken by these blocks. In cases where strict spectral requirements apply, it is not rare to see a large cut of the chip footprint being taken by the reconstruction filter [Oli12]. Another downside of using these bulky filters in a transmitter's signal path, for example, is the bandwidth limitation imposed to higher frequency components intentionally added to the baseband signal in order to compensate for circuit's nonlinearities, reducing the effectiveness of digital pre-distortion (DPD). Anyhow, to any sort of application a DAC implementation that can provide intrinsic noise filtering capabilities along with the already discussed alias attenuation would be very much welcome.

One obvious implementation of the charge-based DAC would comprise a current DAC, where the current delivered to the output load (I_{OUT}) would be readjusted

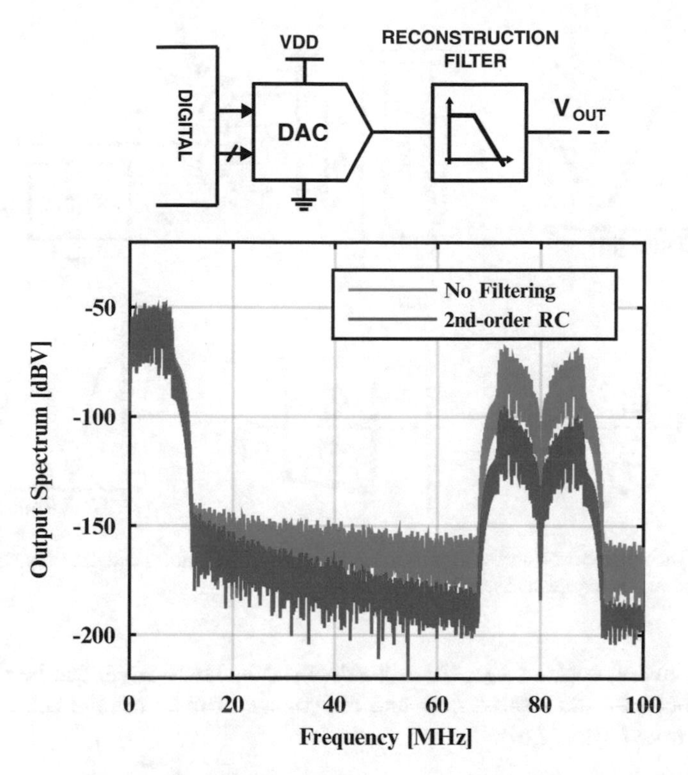

Fig. 2.14 Noise and Alias filtering achieved with a second-order reconstruction RC filter with a cutoff frequency of 15 MHz

every sampling period in order to provide the wanted $Q/\Delta T_S$, as defined by the charge calculation block. In this way, it is clear that the required $Q_{OUT}[k]$ would be uniformly distributed over time, thus leading to the aforementioned $\text{sinc}(\text{x})^2$ alias attenuation.

$$Q_{OUT}[k] = (V_{IN}[k] - V_{IN}[k-1]) \cdot C_{LOAD}$$

$$I_{OUT}[k] = \frac{Q_{OUT}[k]}{T_S}$$

The same functionality can also be achieved with an alternative implementation where the QDAC is supplied with a fixed voltage source, and the same I_{OUT} is provided with a variable conductance that is changed over time according to the digital input command (Fig. 2.15). In the ideal case where the output voltage is an

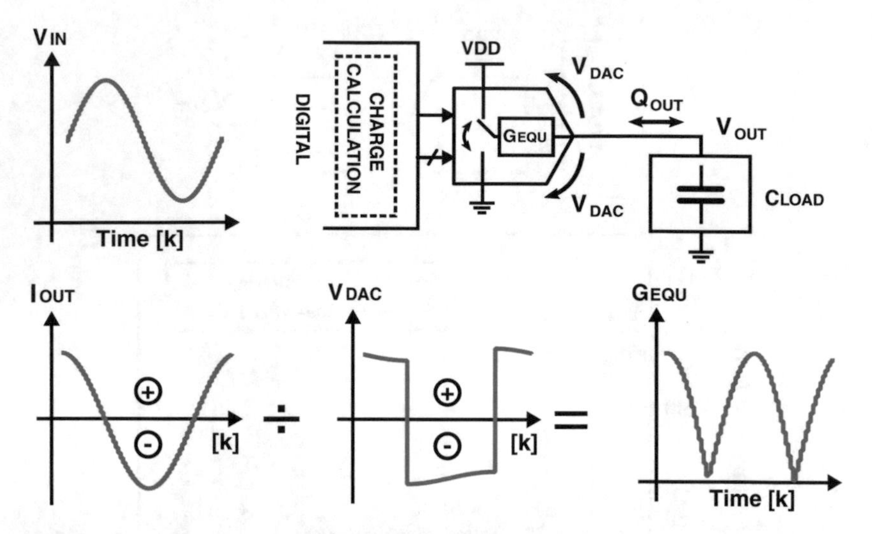

Fig. 2.15 The voltage drop across the QDAC terminals in combination with the output current can be translated into an equivalent DAC conductance

unaltered analog copy of V_{IN}, the equivalent conductance G_{EQU} can be derived as the ratio between the wanted I_{OUT} and the voltage drop across the DAC terminals (V_{DAC}) at time k [Eq. (2.6)].

$$V_{DAC}[k] = \begin{cases} V_{DD} - V_{IN}[k-1], & \text{if } Q_{OUT}[k] > 0 \\ G_{ND} - V_{IN}[k-1], & \text{if } Q_{OUT}[k] < 0 \end{cases}$$
$$G_{EQU}[k] = \frac{I_{OUT}[k]}{V_{DAC}[k]} \tag{2.6}$$

Notably, the combination between the equivalent DAC conductance (G_{EQU}) and the capacitive component of the output load creates an intrinsic single order low-pass filtering effect. What is very particular however, is the fact that the "resistive" component of this RC filter is signal dependent. When a large difference between two consecutive digital input samples is observed (large $\Delta V_{IN}/\Delta T_S$), the increased Q_{OUT} required leads to an also large equivalent G_{EQU} and hence small RC time constant. On the other hand, when a small amount of charge is required (small $\Delta V_{IN}/\Delta T_S$), the instantaneous RC constant is increased (Fig. 2.16). Therefore, from the signal point-of-view no attenuation is implied by the time-varying RC filter, since its resistive component is automatically adjusted according to the wanted output swing.

On the other hand, noise sources which are not correlated to the input signal do not observe the same "on-demand" bandwidth adjustment, and as a result they are filtered by the structure. As demonstrated in Fig. 2.17, noise contributors (such as supply-coupled and DAC-generated noise—including quantization) are intrinsically

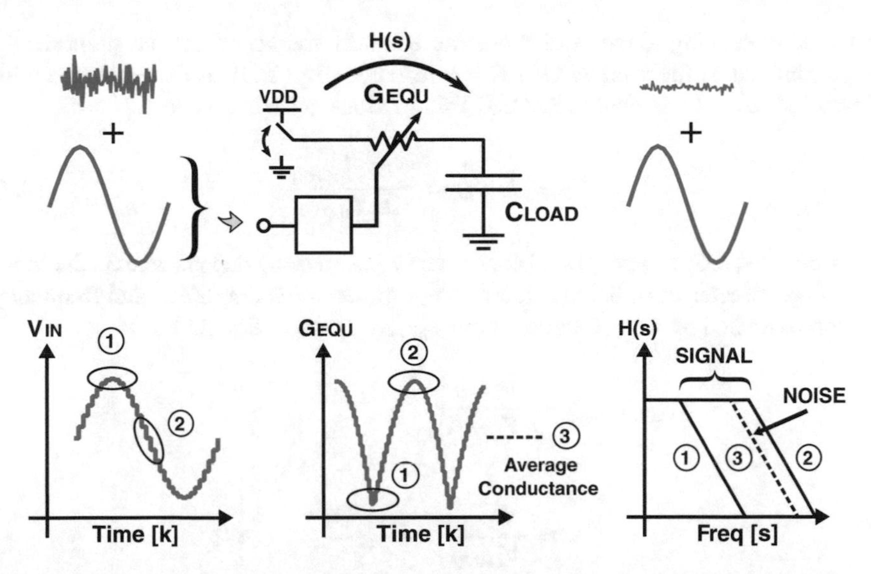

Fig. 2.16 Intrinsic noise filtering mechanism. The signal dependence of the DAC G_{EQU} makes sure that both slow (*1*) and fast (*2*) transitions of the desired signal are properly constructed at the output, facing no attenuation. Uncorrelated noise, on the other hand, is filtered by an equivalent cutoff frequency given by the average conductance (*3*)

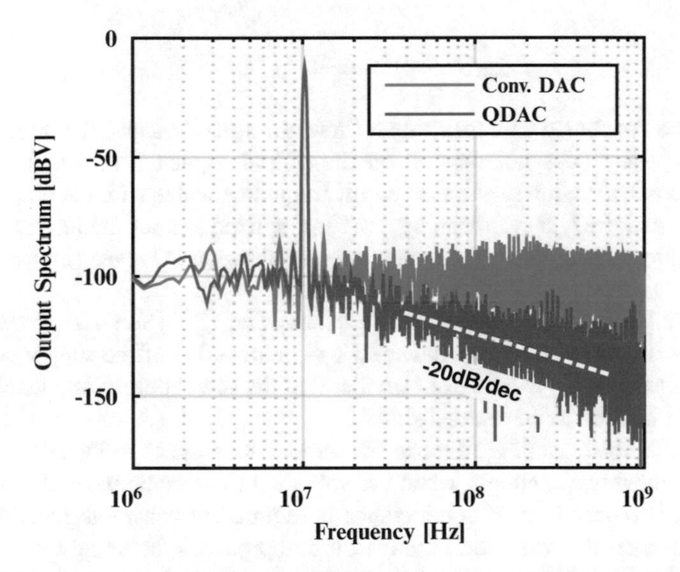

Fig. 2.17 Single-order intrinsic noise filtering, showing a 0.5 V_{pk-pk} 10 MHz single-tone

filtered. Interestingly, the equivalent noise cutoff frequency can be numerically approximated by the average QDAC conductance [Eq. (2.7)], as demonstrated with Periodic Steady-State Noise (PNOISE) simulations in Chaps. 3 and 4.

$$f_{-3\mathrm{dB}}(Noise) = \frac{G_{EQU}|_{Average}}{2\pi C_{LOAD}} \tag{2.7}$$

Moreover, the average QDAC conductance has a strong dependence on the input signal's characteristics, being directly proportional to its amplitude and frequency as approximated by the derivative of the V_{IN}, as shown in Eq. (2.8).

$$\begin{aligned} G_{EQU}|_{Average} &= \left.\frac{I_{OUT}[k]}{V_{DAC}[k]}\right|_{Average} \\ &= \left.\frac{C_{LOAD}}{V_{DAC}[k]} \cdot \frac{\Delta V_{IN}[k]}{T_S}\right|_{Average} \quad \left(\frac{dV_{IN}}{dt}\right) \\ &\propto C_{LOAD} \cdot A \cdot \omega \end{aligned} \tag{2.8}$$

Thus,

$$f_{-3\mathrm{dB}}(Noise) \propto A \cdot \omega \tag{2.9}$$

It can be concluded as a result that: First, for a purely capacitive load the noise cutoff frequency does not depend on the output capacitance, nor the sampling speed. Second, the equivalent noise cutoff frequency scales with the output signal's amplitude and frequency, meaning that for a fixed output frequency the noise cutoff frequency is divided by 2 when the output swing is halved (improving noise filtering).

Figure 2.18 shows the noise cutoff frequency [Eq. (2.7)] versus output frequency for various amplitudes. As noted, for swings below 90 % of the supply voltage the noise cutoff frequency can be even smaller that the actual output frequency, with no attenuation implied to the wanted signal.

The same noise scaling effect is observed with respect to signal's amplitude, with only one remark though: when the voltage difference between the supply and the output is reduced, more conductance is required to convey a given amount of charge. As a result, when the output swing approaches the supply rails (roughly above 70 % of the supply voltage) the DAC average conductance starts increasing exponentially, followed by a corresponding increase of the noise cutoff frequency as observed in Fig. 2.19.

Again, these assumptions are all validated on Chaps. 3 and 4 with PNOISE simulations using the actual QDAC implementations. Although a bit counter-intuitive when first considered, this remarkable feature of a time-varying cutoff frequency—that automatically scales with signal frequency and signal amplitude—is a major advantage of the incremental-charge-based architecture.

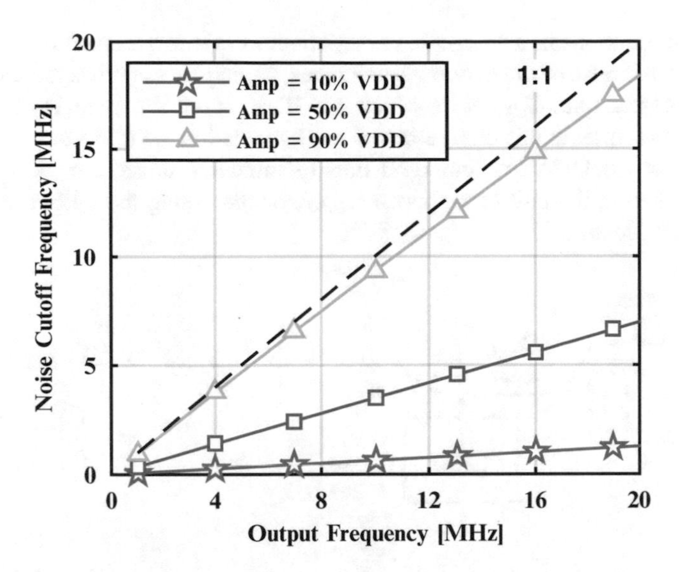

Fig. 2.18 Noise cutoff frequency versus signal output frequency

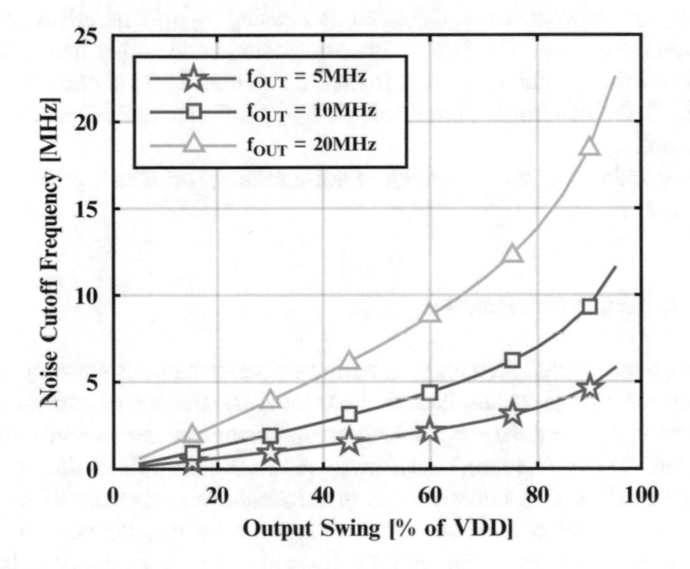

Fig. 2.19 Noise cutoff frequency versus signal amplitude

2.2.1.3 Quantization Noise Scaling

Besides noise filtering given by the signal-dependent RC filter mentioned in Sect. 2.2.1.2, depending on the actual QDAC implementation the quantization noise can be even further reduced.

Chapter 3 introduces a charge-based architecture where quantization noise can be scaled with the ratio between two capacitances, namely an output capacitance C_{LOAD} and a unit capacitance C_{UNIT} with which the QDAC is implemented. Resembling the watermill example, in this charge-based implementation the minimum voltage step resolvable at the QDAC output (and thus quantization noise) can be reduced by either decreasing the unit capacitance C_{UNIT}, or increasing the output capacitance C_{LOAD}, if applicable.

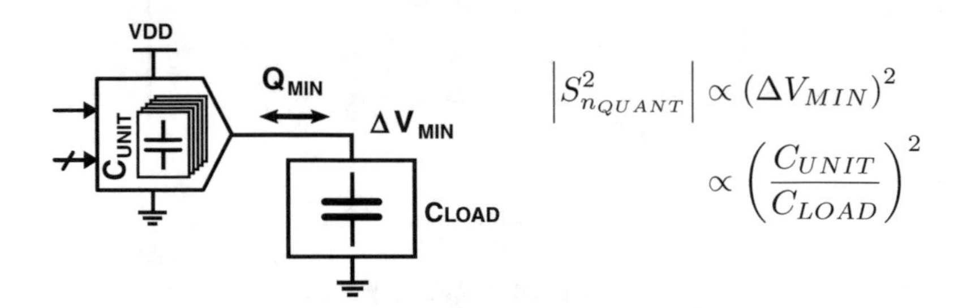

As such, the equivalent number of bits (ENOB) in this architecture can be adjusted by either decreasing C_{UNIT} or increasing C_{LOAD}, based on a required quantization noise level. The DAC number of units, on the other hand, determines what is the maximum amount of charge that can be transferred at once to the output, thus defining the maximum achievable $\Delta V_{OUT}/\Delta T$ that can be produced at the QDAC output.

Further details about the quantization noise scaling feature are given in Chap. 3, Sect. 3.2.3.2.

2.2.1.4 Harmonic Performance

In the charge-based DAC architecture, harmonic performance is strongly dependent on an accurate charge accumulation. Errors in the amount of charge conveyed between the supply and the output load create distortion, hence corrupting signal integrity and affecting harmonic performance. To illustrate the point, for instance the charge-based DAC operation takes into account the existence of a capacitive component to the output load on which charge can be accumulated. For a purely capacitive load, for instance, the output voltage can be represented as follows:

$$\begin{aligned} V_{OUT}[k] &= V_{OUT}[k-1] + \frac{Q_{OUT}[k]}{C_{LOAD}} \\ &= V_{OUT}[k-1] + \frac{(V_{IN}[k] - V_{IN}[k-1]) \cdot C'_{LOAD}}{C_{LOAD}} \end{aligned} \quad (2.10)$$

where C'_{LOAD} is load capacitance value assumed in calculations [Eq. (2.3)].

Since the output voltage is expected to be an unaltered (analog) copy of V_{IN}, an eventual discrepancy between the assumed load capacitance (C'_{LOAD}) and its actual value (C_{LOAD}) can be accounted as:

$$V_{OUT}[k] = V_{IN}[k] + \text{error}[k] \tag{2.11}$$

where the error at time k is given by:

$$\begin{aligned}\text{error}[k] &= V_{IN}[k] \cdot \left(\frac{C'_{LOAD}}{C_{LOAD}} - 1 \right) \\ &\quad + (V_{OUT}[k-1] - V_{IN}[k-1]) \cdot \frac{C'_{LOAD}}{C_{LOAD}} \\ &= V_{IN}[k] \cdot \left(\frac{C'_{LOAD}}{C_{LOAD}} - 1 \right) + \text{error}[k-1] \cdot \frac{C'_{LOAD}}{C_{LOAD}} \end{aligned} \tag{2.12}$$

The analysis of Eq. (2.12) reveals two important aspects: First, a correct charge-based operation relies on a good knowledge of the output load. The more distant the assumed load capacitance (C'_{LOAD}) is from the its actual value (C_{LOAD}), the larger the error produced. Second, to make matters worse, the recursive dependence of the error function to its previous value (error$[k-1]$) demonstrates that the memory effect introduced by the output capacitance not only integrates the wanted charge, but also the errors implied in the charge accumulation. These errors can be caused by multiple reasons, from calculation mistakes to implementation imperfections—including charge leakage.

To illustrate the severeness of error accumulation in the charge-based architecture, Fig. 2.20 shows the impact of a one-time only event, where the amount of charge transferred to the load Z at a particular sampling period is mistakenly reduced by 10 %. If the subsequent charge blocks are kept unchanged, the system operation would be quickly compromised since the accumulated error would make the output voltage drift to one of the supply rails.

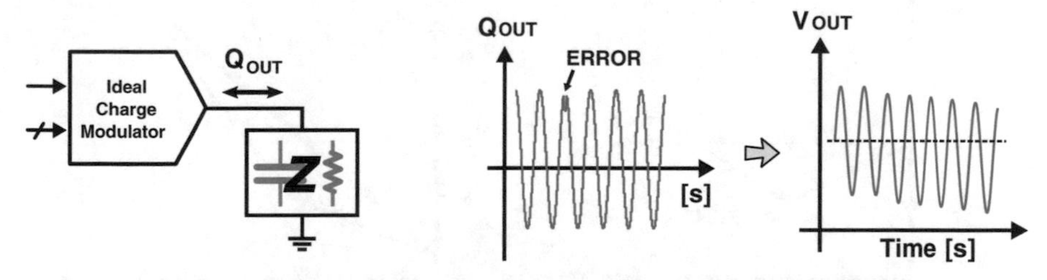

Fig. 2.20 Error accumulation can quickly compromise the charge-based operation if subsequent blocks are not re-adjusted to account for a wrong packet of charge

Basically, the charge-based operation simply would not be feasible if it was not for the "self-healing" effect introduced by the fact that the output charge is also dependent on $\Delta V's$. In the proposed architecture, the amount of charge transferred to Z during a given sampling period is always proportional to the voltage difference between the supply voltage and the previous V_{OUT}. If an actual time-varying G_{EQU} would be connected between the supply and the output load in order to control the charge output, the amount of charge conveyed to Z in this case would be:

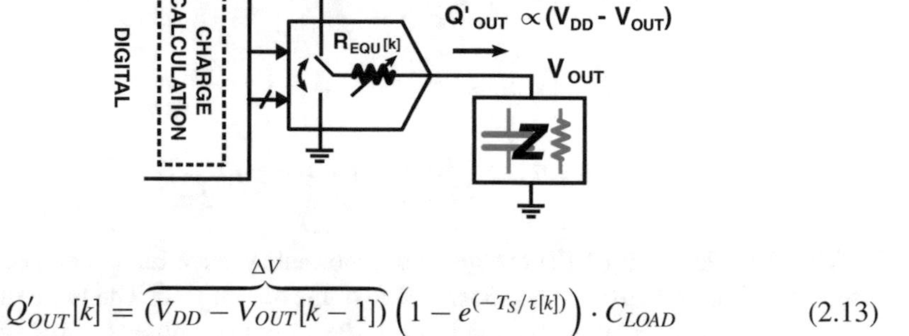

$$Q'_{OUT}[k] = \overbrace{(V_{DD} - V_{OUT}[k-1])}^{\Delta V} \left(1 - e^{(-T_S/\tau[k])}\right) \cdot C_{LOAD} \tag{2.13}$$

where $\tau[k] = R_{EQU}[k] \cdot C_{LOAD}$ and $T_S = 1/F_S$.

The ΔV dependence functions here as a intrinsic feedback where Q'_{OUT} is automatically adjusted when $V_{OUT}[k-1]$ deviates from the expected $V_{IN}[k-1]$, avoiding thus the infinite charge error accumulation.

The principle works as follows: Suppose that equal amounts of charge were to be transferred to the output load at both times A and B ($Q_{A,B}$), but the first charge "block" (Q'_A) is unexpectedly reduced, making the output voltage reach V'_A instead of the wanted V_A. As shown in Fig. 2.21, the increased voltage difference between the "wrong" V'_A and the supply voltage V_{DD} will act in this case increasing the amount of charge transferred at time B (Q'_B).

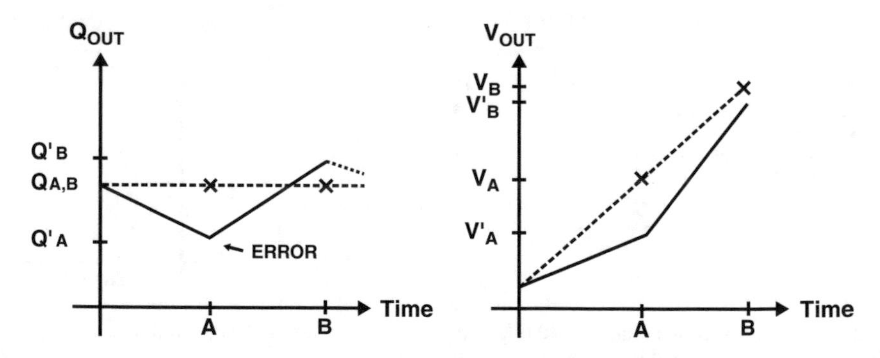

Fig. 2.21 "Self-healing" mechanism. The increased voltage difference implied by a reduced charge "block" increases the amount of charge transferred during the following steps

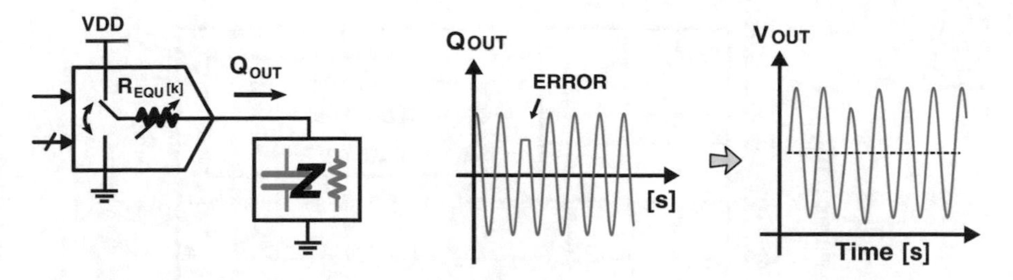

Fig. 2.22 Thanks to the "self-healing" mechanism, the charge-based operation is completely resilient even to large errors in the charge accumulation

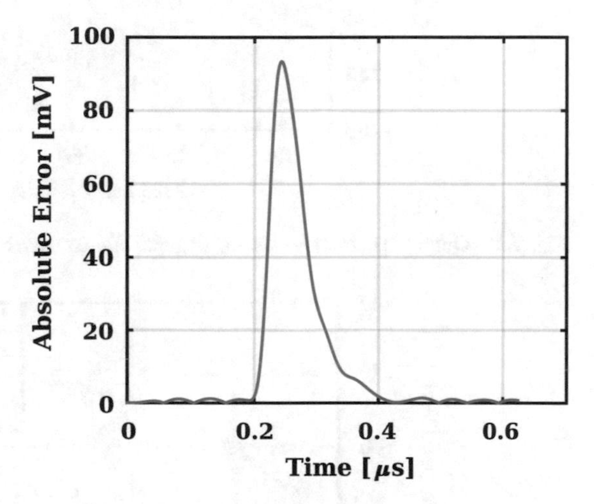

Fig. 2.23 Even when large errors are implied, the accumulated error is quickly dissipated

The improved resilience to error accumulation is shown in Fig. 2.22 for a 10 MHz output single-tone, where a total of 50 charge blocks are clipped to a single value representing almost 40 % reduction in the total amount of charge required during that period. As noted, only a localized swing reduction is observed, which is quickly restored once the disturbance is removed.

The architecture's fast error "dissipation" capability can also be seen in Fig. 2.23. Even in the impracticable situation of 40 % charge reduction, the output voltage error falls to a minimum in less than two cycles.

Though error accumulation in the proposed architecture does not prevent a correct operation, it still affects the charge-based architecture by introducing signal distortion and harmonic degradation. While specific details of each proposed implementation are shown in Chaps. 3 and 4, a broad analysis of charge-based architecture's harmonic performance is performed here.

Figure 2.24 shows the output spectrum for several degrees of mismatch between the actual output load capacitance (C_{LOAD}) and its assumed value (C'_{LOAD}). As noticed, a clear impact in the odd harmonics is observed, leading to the conclusion (based on Fig. 2.25) that, in order to achieve a third-order harmonic distortion below −60 dBc, an accuracy better than 2 % is necessary.

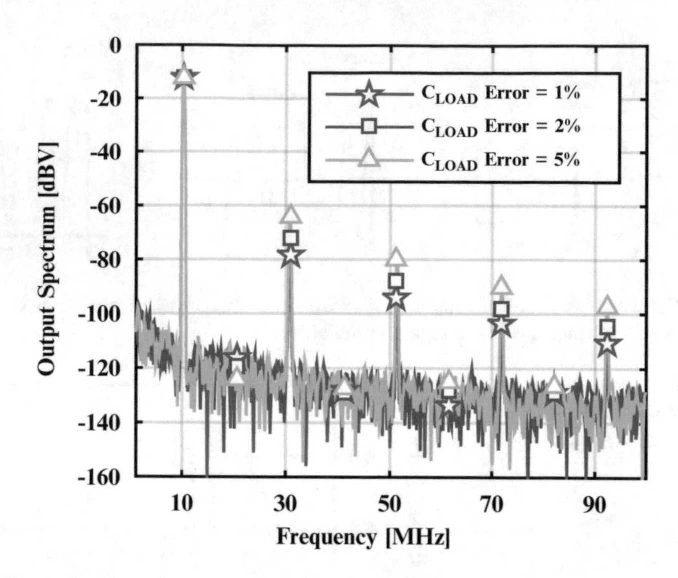

Fig. 2.24 Output Spectrum for several degrees of mismatch between C_{LOAD} and its assumed value

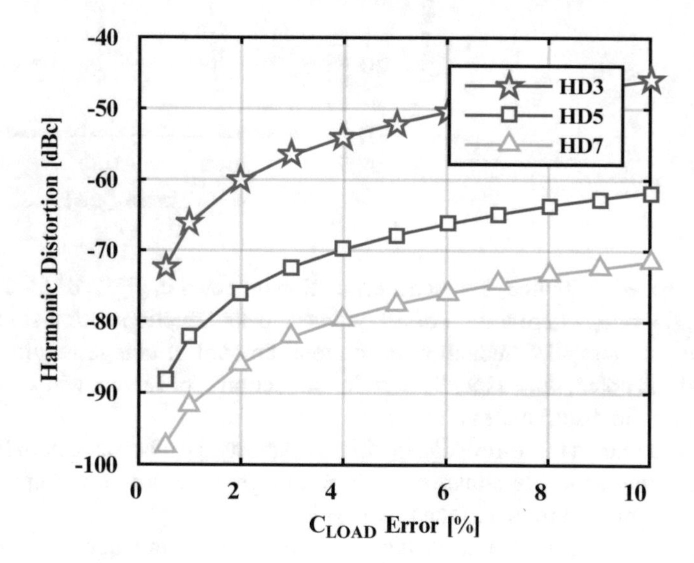

Fig. 2.25 Harmonic distortion versus C_{LOAD} mismatch

Figure 2.26 demonstrates the impact of gain error in the output spectrum. This situation applies when the QDAC unit cell deviates from its expected value. Similarly, a large mismatch incurs increased harmonic distortion, and again less than 2 % mismatch is required to achieve an improved harmonic performance.

Though achieving such degree of accuracy may seem challenging at first, each one of these parameters can be calibrated after fabrication, using low frequency

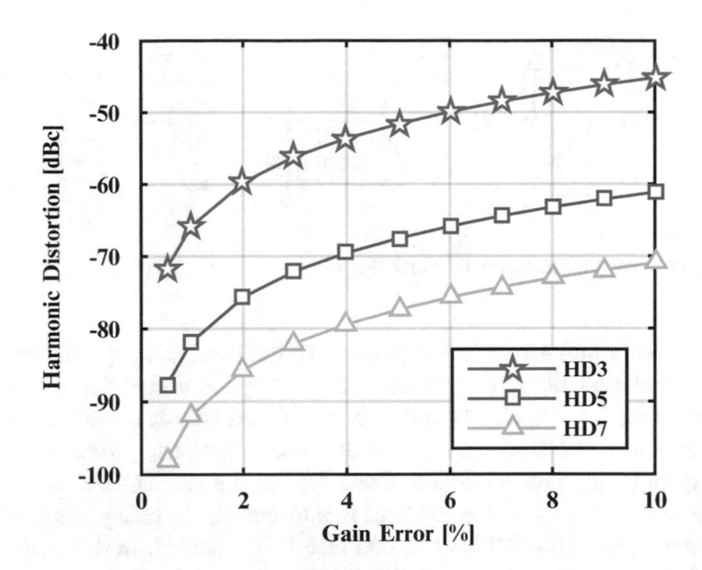

Fig. 2.26 Harmonic distortion versus QDAC gain mismatch

measurements to deduct their exact values. Instantaneous variations of one or more load impedances (such as antenna impedance variations due to changing surrounding conditions) will necessarily require a feedback loop to change the corresponding parameters involved in the charge calculation. Nevertheless, further improvements in harmonic performance can also be targeted by improving the algorithm to account for circuit non-idealities.

2.3 Charge-Based Transmitter

In light of all noise filtering characteristics discussed in Sect. 2.2.1, it seems logical to apply the charge-based operation in order to achieve the stringent noise specifications required to SAW-less transmitters in full duplex.

Starting from a conventional architecture, a typical transmitter implementation is composed of a DAC, followed by a reconstruction filter and a mixer to perform the frequency translation. In most cases, a Pre-Power Amplifier (PPA) is also added as an output stage before driving the PA. In the proposed transmitter architecture, the DAC and the reconstruction filter are exchanged for a QDAC and a baseband capacitance, all the rest remaining the same (Fig. 2.27). Instead of consuming excessive DC power in order to drive the capacitive nodes using strong voltage buffers, now both baseband and RF nodes are going to be driven with charge. Therefore, in the following analysis, the RF impedance is represented as a generic load Z.

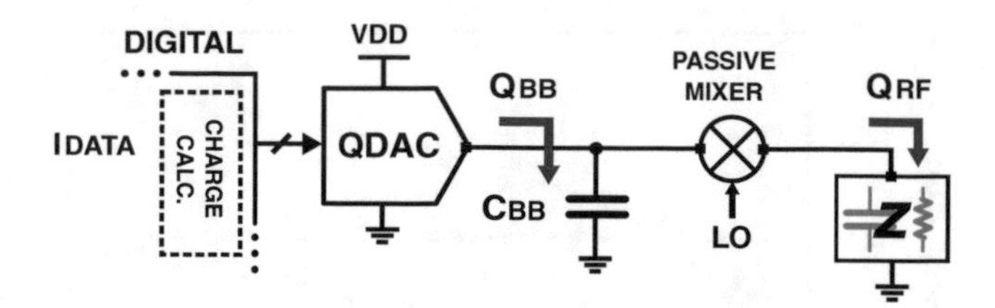

Fig. 2.27 Simplified charge-based TX block diagram

As in the watermill example, the purpose of the baseband capacitance C_{BB} is to provide the functionality of a reservoir, where charge is first buffered before being transferred to the RF output. In combination with the QDAC operation, this charge reservoir should provide all the benefits discussed previously, from intrinsic noise filtering to sampling alias reduction. Compared to the basic QDAC operation, the difference now is that once the RF load is introduced, the charge subtracted from C_{BB} in order to drive the RF load should also be accounted. In this case, the total amount of charge per sampling period required to operate the charge-based structure becomes:

$$Q_{TOTAL} = Q_{BB} + Q_{RF} \tag{2.14}$$

where Q_{BB} corresponds to the amount of charge required to move the baseband voltage across the consecutive input samples, and Q_{RF} stands for the amount of charge subtracted from C_{BB} while driving the RF load Z every time the passive mixer switch is closed.

The operating principles are kept the same. To prove the functionality an ideal passive mixer is added, with a load impedance Z equal to the input capacitance that would be expected from a PPA implementation with a compression point around 10 dBm. Using the same "black-box" QDAC implementation where the output charge Q_{TOTAL} is uniformly distributed over the sampling period, Fig. 2.28 shows the output spectrum of an example 20 MHz multi-tone baseband signal transmitted at 2 GHz. As expected, both noise filtering capabilities and alias attenuation are also clearly noticeable at LO frequency.

The inclusion of a passive mixer and a RF load in the architecture has mainly two effects: First, the total amount of charge consumed (Q_{TOTAL}) is largely increased when a low impedance RF load is used, requiring a similar increase in the QDAC charge capacity. In the two implementations proposed in this work, in both cases the charge capacity is increased by increasing the total number of QDAC elements that can be used to convey charge to the baseband capacitance. In the watermill analogy, it corresponds to increasing the number of buckets available.

Second, since the passive mixer does not provide isolation between baseband and RF nodes, the noise cutoff frequency is also affected by the equivalent RF load impedance translated to the baseband side. As also discussed in [Raz12], when looking from the baseband node at frequencies much smaller than the mixer

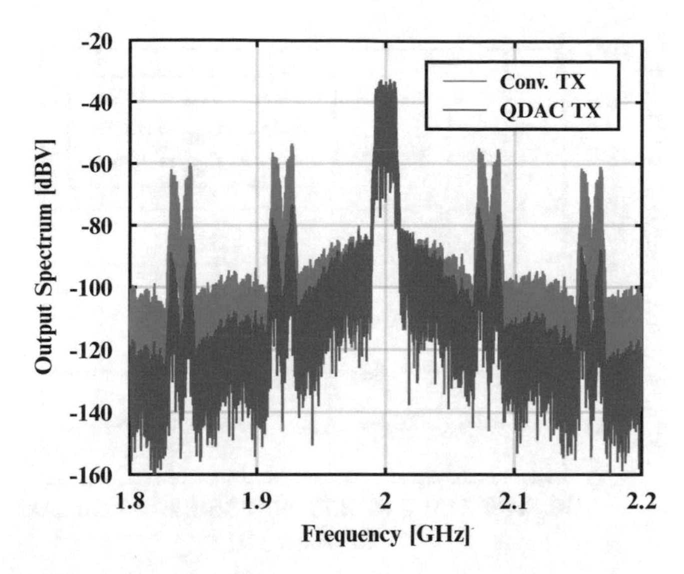

Fig. 2.28 RF spectrum of a charge-based TX driving a 200 fF capacitance at 2 GHz

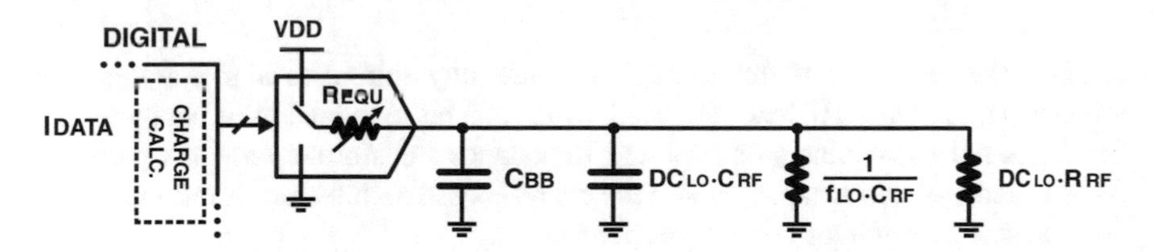

Fig. 2.29 Simplified schematic where the RF load is translated to the baseband side

switching frequency ($\omega << 2\pi f_{LO}$), an output RF capacitance can be translated into two components: A capacitive component given by C_{RF} scaled by the LO duty cycle (DC_{LO}), and a resistive component given by the average (DC) current drained by C_{RF} ($R_{C_{RF}} = 1/C_{RF}f_{LO}$). Similarly, an RF resistance is also translated to the baseband by looking at the average current drained by R_{RF}. In a differential implementation where each baseband node is connected twice per LO period to the RF node, the LO duty cycle is doubled. Figure 2.29 shows the equivalent schematic considering the translated RF load.

The resulting noise cutoff frequency considering the RF load is hence given by:

$$f_{-3dB}(Noise) = \frac{\left(G_{EQU}|_{Average} + f_{LO}C_{RF} + \dfrac{1}{DC_{LO}R_{RF}} \right)}{2\pi(C_{LOAD} + DC_{LO}C_{RF})} \tag{2.15}$$

Figure 2.30 shows the impact of a resistive RF load for various baseband capacitances considering a 0.7 Vpp 10 MHz single-tone transmitted at 2 GHz. As

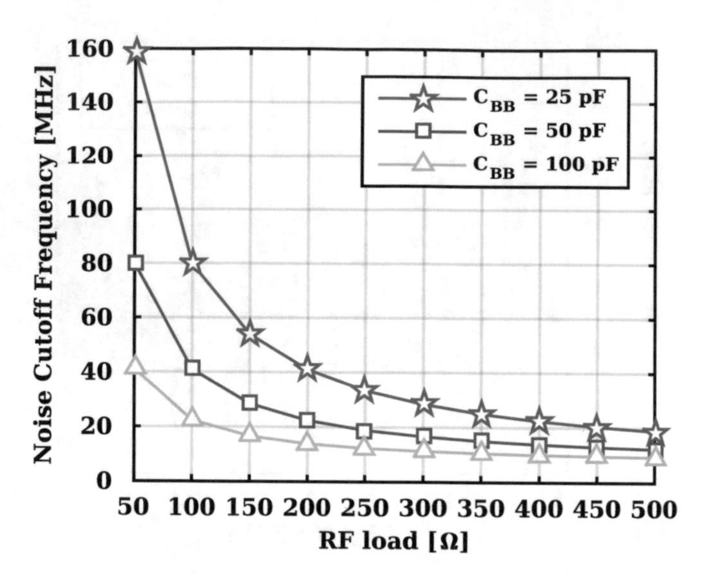

Fig. 2.30 Noise cutoff frequency versus RF load resistance

noticed, the noise cutoff frequency is considerably shifted to higher frequencies when low impedance RF loads are used, which can be addressed by either increasing the baseband capacitance, or applying impedance transformation. This particular issue is addressed in Chap. 4, where the charge-based architecture is used to drive a 50 Ω load representing the PA input impedance.

2.3.1 Power Efficiency

Besides noise reduction, incremental charge-based operation can also provide great improvements in terms of power efficiency. For a more fundamental understanding of the potential improvements and how it trades with noise performance, only the charge intake represented by Q_{TOTAL} is considered in this analysis. The impact of additional contributors including QDAC digital operation and LO buffering are discussed in subsequent chapters.

By first looking at QDAC independently and considering the amount of power required to drive the baseband capacitance alone, it can be noticed that the charge-based architecture fundamentally behaves as a Class-B topology, since it only drains charge from the supply when the total amount of charge in the system should be increased—thus for only half of the signal period (Fig. 2.31). As a result, when driving the baseband capacitance only, the charge-based DAC can provide a maximum efficiency of 78.2 % as verified in simulations and shown in Fig. 2.32 for various signal amplitudes. However, the maximum efficiency corresponds to an output swing equal to V_{DD}, which cannot be realized since it would require an

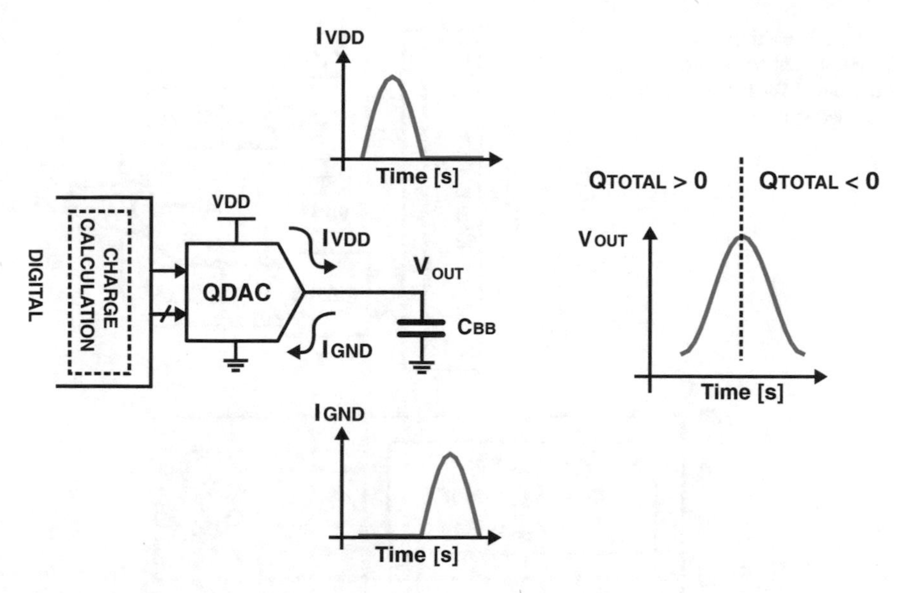

Fig. 2.31 Simplified schematic depicting QDAC Class-B operation

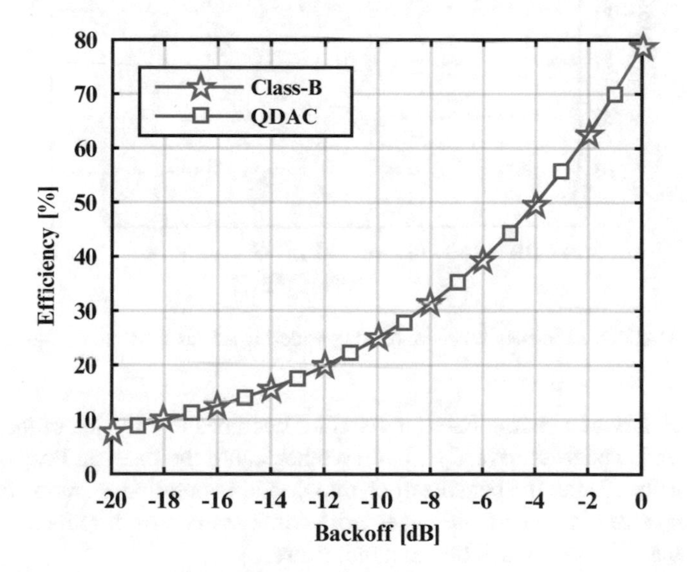

Fig. 2.32 Efficiency versus Backoff considering C_{BB} only

infinite QDAC conductance. Nevertheless, swings of at least 90 % of the supply voltage are completely feasible, corresponding to a QDAC efficiency of 70 %.

When the RF load is included, the power consumption becomes a function of two contributors, namely the baseband and RF charges. Since the Q_{RF} component corresponds to the only fraction of Q_{TOTAL} that is actually transmitted, the maximum

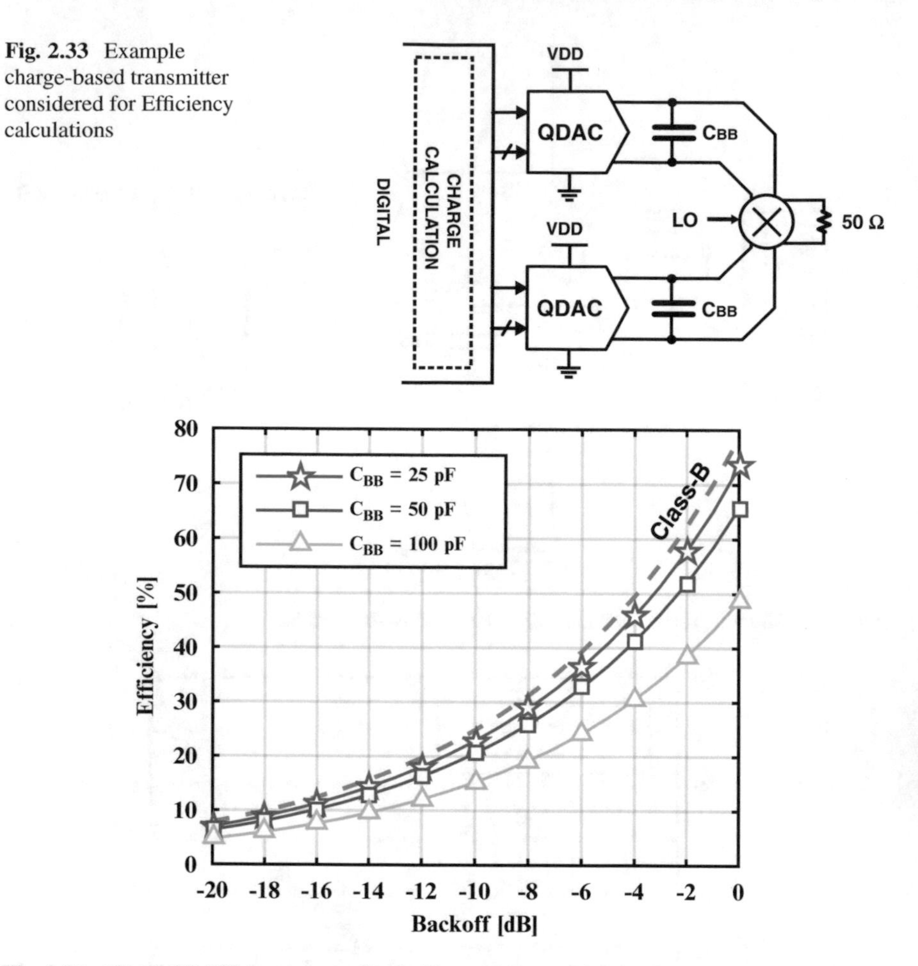

Fig. 2.33 Example charge-based transmitter considered for Efficiency calculations

Fig. 2.34 QDAC TX Efficiency versus Backoff considering a 50 Ω load and various C_{BB} values

power efficiency of a charge-based transmitter becomes mainly determined by how much power is spent to drive C_{BB}. On the other hand, the baseband capacitance is also responsible for all the benefits of charge-domain operation, so reducing C_{BB} (in order to improve efficiency) will necessarily come at the cost of reducing the noise filtering capabilities provided by the architecture.

This tradeoff is explored here with a simple exercise considering a differential I/Q charge-based TX driving a 50 Ω load (Fig. 2.33). In this example, the efficiency versus backoff from theoretical maximum swing ($V_{pp} = V_{DD}$) is shown in Fig. 2.34. As noticed, the increasing amount of power spent to drive the baseband capacitances decrease the maximum achievable power efficiency as C_{BB} is increased.

To demonstrate how efficiency trades with noise filtering in this specific case, Fig. 2.35 shows both efficiency and noise cutoff frequency at 2 dB backoff for different baseband capacitances.

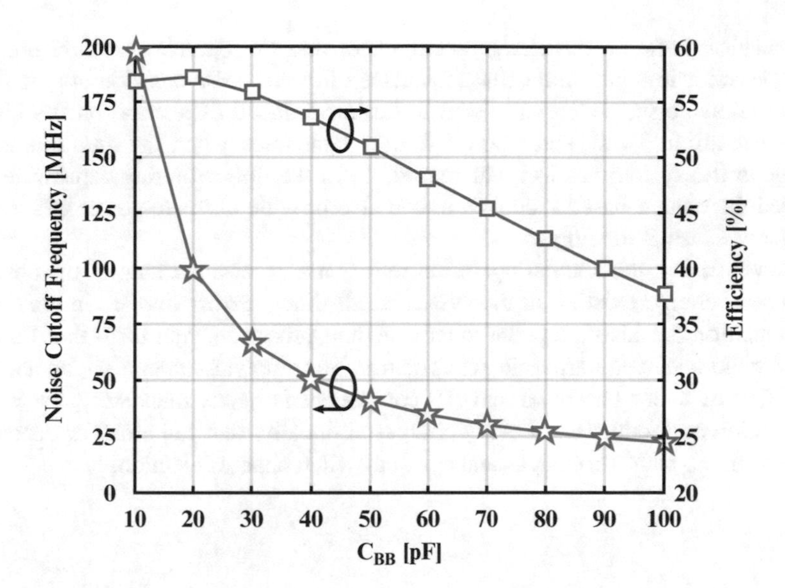

Fig. 2.35 Efficiency and noise cutoff frequency versus C_{BB} at 2 dB backoff

Thus, the charge-based architecture provides a flexible solution where power efficiency can be easily improved in cases where OOB noise requirements are more relaxed, which unfortunately is not the case in our target application. Therefore, the following chapters make a solid case demonstrating the noise filtering capabilities, rather than the considerable improvements in power efficiency that could also be attained with the charge-based architecture.

2.4 Conclusion

In this chapter the fundamentals of incremental-charge-based operation are discussed. Starting with the analogy of a watermill whose output power has to be controlled in a precise and timely fashion, most of the benefits derived from this unconventional operating mode could be implied. The incremental-charge-based operation is based on a controlled charge convey and accumulation at the different nodes existing in the system. The voltage swings are produced by "incrementally" adding or subtracting charge from the several charge accumulators, in a way that only the added charge is drained from the supply.

Using a generic QDAC "black-box" implementation that is capable of delivering controlled amounts of charge defined with simple calculations, three main advantages of incremental-charge-based operation could be demonstrated: First, the continuous (or discrete-time) charge accumulation yields an inherent quasi-linear interpolation that significantly attenuates the sampling aliases. Second, the combination between the charge accumulator and the QDAC operation provides a single-order low pass filter that attenuates different noise contributors including quantization noise, which can be even further reduced depending on the QDAC implementation. Third, since power is only consumed when the total amount of charge in the system has to be increased, when the noise filtering capabilities are relaxed the charge-based architecture can also provide ultimate power efficiencies similar to Class-B topologies.

However, the incremental operation relies on a precise tracking of the absolute amount of charge existing in the system at all times. Errors implied in the charge accumulation are also integrated over time and produce signal distortion that can degrade the achieved harmonic performance. For a flawless operation, an accurate definition of every baseband and RF component is hence mandatory, as well as observation and reduction of every circuit non-ideality that can affect or distort the system charge balance, such as leakage and switch charge-injection.

Chapter 3
Capacitive Charge-Based Transmitter

3.1 Introduction

In light of all the potential noise and efficiency benefits discussed in Chap. 2, the first proof-of-concept charge-based transmitter implementation [Par15b, Par15a] is presented in this chapter.

As it will be noted in the following paragraphs, this first implementation closely resembles the proposed watermill analogy, where the amount of power transferred to an output load is controlled by either increasing or decreasing the water level in a reservoir. Instead of buckets and valves however, switches and capacitors are used in this case to deliver controllable amounts of charge to a baseband capacitor.

The motivations for choosing this first topology are manifold: First, an architecture exclusively based on switches and capacitors is inherently "digital friendly", in the sense that it can be easily scaled and ported to different technologies following the typically dominant digital circuitry. Second, switches and capacitors are perhaps the only two components that benefit from scaling in advanced digital-oriented CMOS technologies. By making smaller transistors the intrinsic capacitances are reduced, allowing faster switching with less power consumption [Raz12]. The feature size reduction also provides larger capacitance densities, improving area efficiency with metal-oxide-metal (MOM) integrated capacitors. Third, and not less important, as further discussed in Sect. 3.2.3.2 the given topology also provides quantization noise scaling capabilities, relaxing the out-of-band noise emission.

This chapter is divided into other four sections. Section 3.2 provides the architecture description with operating principles, followed by circuit implementation (Sect. 3.3) and layout considerations. Measurement results and conclusions are given in Sects. 3.4 and 3.5 respectively.

P.E. Paro Filho et al., *Charge-based CMOS Digital RF Transmitters*, Analog Circuits and Signal Processing, DOI 10.1007/978-3-319-45787-1_3

3.2 Architecture

The first proposed charge-based architecture is inspired in a conventional quadrature direct-conversion IQ transmitter implementation, comprising a digital-to-analog converter (DAC) followed by a reconstruction filter, a mixer and a pre-power amplifier. However, instead of using a power-hungry low-output-impedance DAC output-stage to provide the required bandwidth and linearity, the architecture will now be operated in charge-domain.

An overview of the proposed architecture can be seen in Fig. 3.1. The transmitter architecture consists of four capacitive charge-based DACs (CQDACs), responsible for driving both baseband and RF nodes with discrete packets of charge following the command of a charge calculation block at the baseband engine. A passive mixer is chosen for its improved noise performance and reduced area and power consumptions. Driven by 25 % duty-cycle LO signals, it connects each one of the baseband nodes to PPA input at the appropriate LO phase [He09]. The RF load, in turn, is given by the total capacitance seen at the PPA input. For a compression-point around 10 dBm, the estimated PPA input capacitance is in the range between 200~300 fF.

The CQDAC in turn is implemented through the parallel combination of 1024 unit capacitances that can be individually selected. As shown in Fig. 3.2, it represents a controllable capacitance that can be pre-charged to either V_{DD} or G_{ND}, and connected to the baseband capacitance C_{BB}. At this point the resemblance with the bucket/reservoir example becomes evident.

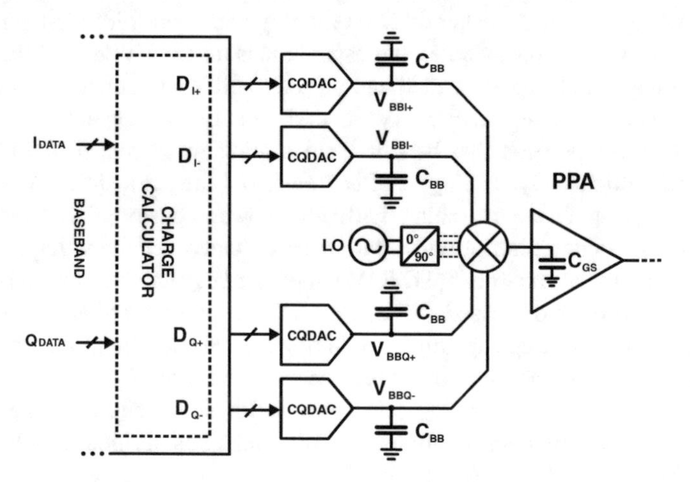

Fig. 3.1 Architecture overview of the direct-conversion IQ charge-based transmitter. Four CQDACs provide each one of the differential I/Q components, driving the pre-power amplifier (PPA) through a passive switch-based mixer

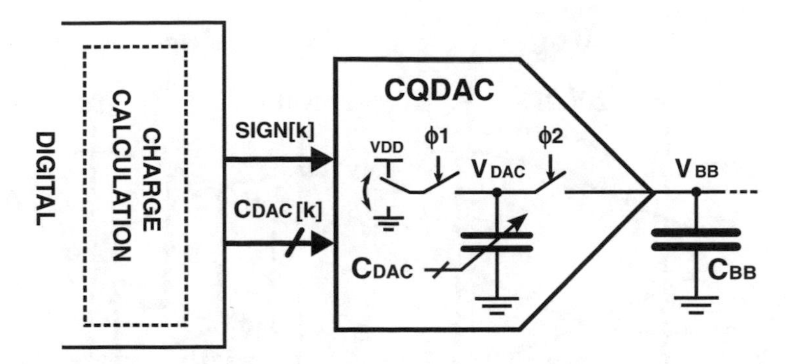

Fig. 3.2 CQDAC block, which delivers discrete packets of charge at LO speed by controlling a variable capacitance $C_{DAC}[k]$, and to what voltage it is pre-charged before connecting to C_{BB}

The charge calculations are performed at baseband speed. The charge calculation block defines the instantaneous number of unit capacitors required to convey the wanted charge, first transmitted to C_{BB}, and later to the RF node.

All charge operations are synchronized on chip using different phases of the LO signal, generated internally and derived from an external differential LO signal.

3.2.1 Operating Principles

To understand how the proposed architecture operates, consider the signal path of a individual I/Q component. In the simplified schematic shown in Fig. 3.3, the mixer is depicted as an ideal switch that closes once per LO cycle at the corresponding phase, loaded with the parasitic capacitance of the PPA input.

Both capacitances C_{BB} and C_{GS-PPA} are going to be driven with discrete packets of charge, chosen in this implementation to be delivered at LO rate. Every baseband sampling period ($1/F_S$), the charge calculation block in the baseband processor determines how much extra charge should be accumulated on (or subtracted from) each capacitor, so that every node can follow its expected voltage excursion.

The total amount of charge required (Q_{TOTAL}) is subdivided into two components: a baseband (Q_{BB}) and a RF component (Q_{RF}). The baseband component corresponds to the necessary charge to move the baseband voltage (V_{BB}) across two consecutive baseband samples ($V_{BB}[k-1]$ and $V_{BB}[k]$). Using the positive in-phase (V_{BB_I+}) I/Q signal as an example, Q_{BB_I+} is hence calculated by:

$$Q_{BB_I+}[k] = (V_{BB_I+}[k] - V_{BB_I+}[k-1]) \cdot C_{BB} \tag{3.1}$$

corresponding to the amount required per F_S cycle.

The RF charge, on the other hand, corresponds to the amount of charge taken by the RF capacitance (C_{GS-PPA}) when moving the RF voltage (V_{RF}) from the previously sampled baseband voltage (V_{BB_Q-}) to the wanted V_{BB_I+}:

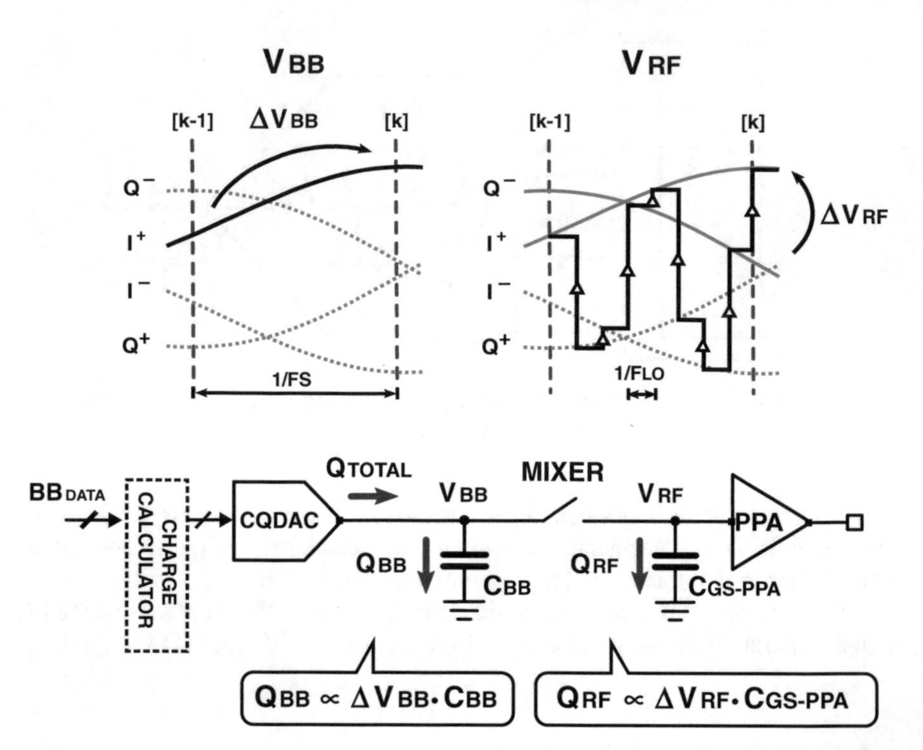

Fig. 3.3 Operating principle of the charge-based transmitter architecture. Based on the digital I/Q input signal, the amount of charge that should be transferred to both baseband and RF nodes are calculated

$$Q_{RF_I+}[k] = \left(V_{BB_I+}[k] - V_{BB_Q-}[k]\right) \cdot C_{GS_{PPA}} \tag{3.2}$$

which is subtracted from C_{BB} every time the mixer switch is closed, thus every LO cycle.

The total charge needed per LO period is the sum of the two charge components, with Q_{BB_I+} scaled by the ratio between the baseband (F_S) and LO frequencies (F_{LO}):

$$Q_{TOTAL_I+}[k] = Q_{BB_I+}[k] \cdot \frac{F_S}{F_{LO}} + Q_{RF_I+}[k] \tag{3.3}$$

It is important to note that even though Q_{TOTAL} is delivered at LO speed, the calculation engine works at baseband F_S frequency, allowing significant reductions in processing power consumption, and interface speed toward the CQDAC.

The minimum speed at which charge calculations should be performed is determined by the targeted harmonic performance. For instance, the assumption that Q_{BB} can be divided into identical F_{LO}/F_S sub-packets [Eq. (3.3)] becomes more accurate when ΔV_{BB} between samples is reduced, requiring a minimally appropriate sampling frequency.

Further, when calculating the RF charge, it can be considered the fact that V_{BB} will be changing from $V_{BB}[k-1]$ to $V_{BB}[k]$ during the given sampling period. Therefore, more accurate results can be achieved in this case by considering the average value between samples when calculating Q_{RF}:

$$Q_{RF_I+}[k] = \left(\frac{V_{BB_I+}[k] + V_{BB_I+}[k-1]}{2} - \frac{V_{BB_Q-}[k] + V_{BB_Q-}[k-1]}{2}\right) \cdot C_{GS_{PPA}} \tag{3.4}$$

3.2.2 CQDAC Operation

The CQDAC transmitter operation is based on a 3-phase charge convey from the supply to the RF load (and from RF load to ground), starting from a controllable capacitance C_{DAC} and accumulated at the baseband capacitor C_{BB}.

As shown in Fig. 3.4, at phase one (called *precharge*) the instantaneous pre-calculated DAC capacitance $C_{DAC}[k]$ is pre-charged to either V_{DD} or G_{ND}, following the control signal *SIGN*[*k*]. If *SIGN*[*k*] is equal to logic "0" (zero), $C_{DAC}[k]$ is pre-charged to V_{DD}, and if *SIGN*[*k*] is equal to "1" (one), $C_{DAC}[k]$ is discharged to G_{ND}. At phase two (*share*), once the supply and mixer switches are opened, both DAC

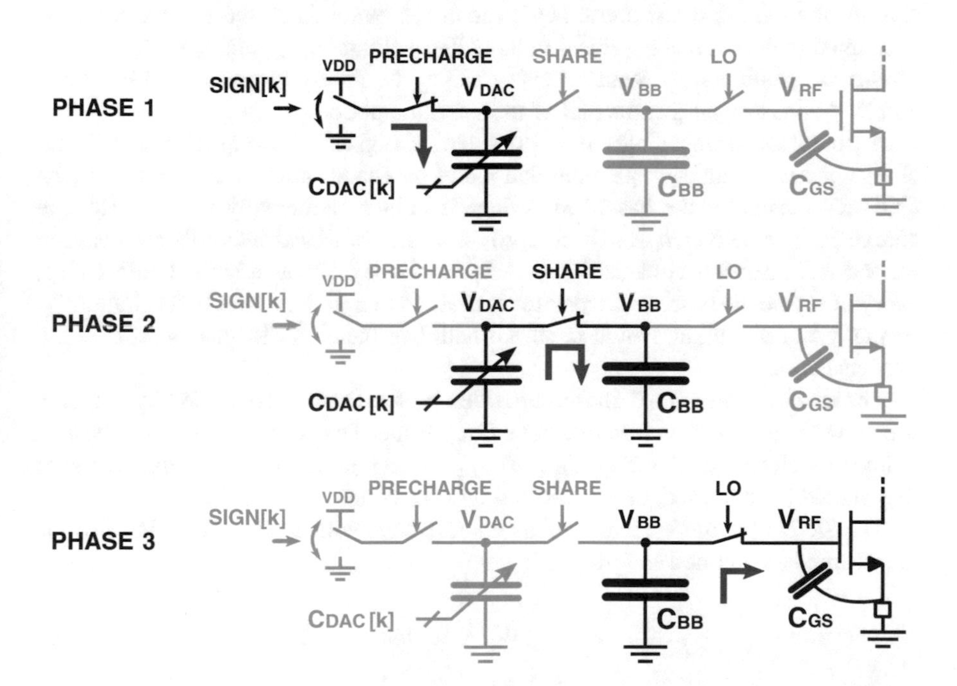

Fig. 3.4 The charge transfer between the supply and the RF node is divided into three non-overlapping phases, namely *precharge*, *share* and *LO*

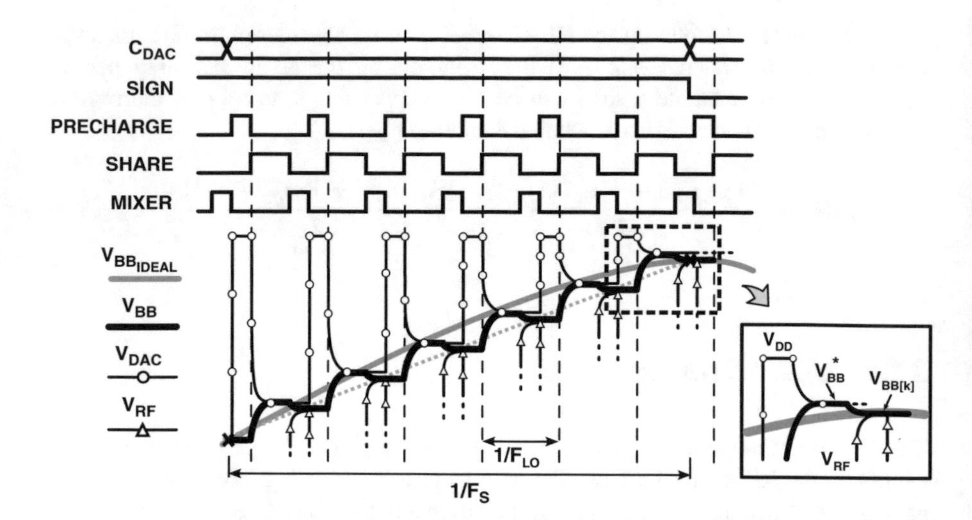

Fig. 3.5 Detailed incremental charge-based operation. Since the amount of charge taken by the RF node is also accounted, the baseband voltage reaches the expected final value $V_{BB}[k]$ at the end of the sampling period

and baseband capacitances are connected together, and $Q_{TOTAL}[k]$ is transferred to C_{BB}. At the third and last phase (*LO*), the mixer switch is closed and the RF node is charged to the wanted baseband voltage V_{BB}, subtracting $Q_{RF}[k]$ from C_{BB}. These phases are continuously repeated every LO cycle, finally bringing both baseband and RF nodes to $V_{BB}[k]$ at the end of the sampling period.

A complete timing diagram is provided in Fig. 3.5 showing all the different charge phases for an example combination of baseband samples. As noticed, in the CQDAC transmitter the baseband voltage is built incrementally with consecutive charge packets delivered at LO rate, providing a quasi-linear interpolation between the wanted baseband voltages $V_{BB}[k-1]$ and $V_{BB}[k]$. By accurately defining Q_{RF}, every LO cycle the baseband capacitor is charged to a slightly modified voltage V^{*}_{BB}, so that when the mixer switch is closed both baseband and RF nodes settle at the expected voltage.

The absolute amount of charge provided by the DAC is controlled by digitally selecting the appropriate instantaneous C_{DAC} value. The charge direction in turn, is defined by either pre-charging C_{DAC} to V_{DD} if the total amount of charge stored at C_{BB} should be increased, or discharging it to G_{ND} otherwise.

The *SIGN* control signal as well as the reference voltage V_{REF} to which C_{DAC} is pre-charged are defined as follows:

$$SIGN_{I+}[k] = \begin{cases} 0, & \text{if } Q_{TOTAL_{I}+}[k] \geq 0, \\ 1, & \text{otherwise.} \end{cases} \quad (3.5)$$

$$V_{REF_I+}[k] = \begin{cases} V_{DD}, & \text{if } SIGN_{I+}[k] = 0, \\ G_{ND}, & \text{otherwise.} \end{cases} \tag{3.6}$$

The instantaneous DAC capacitance, in turn, is calculated considering the charge balance before and after the *share* switch is closed, as demonstrated in Eq. (3.7).

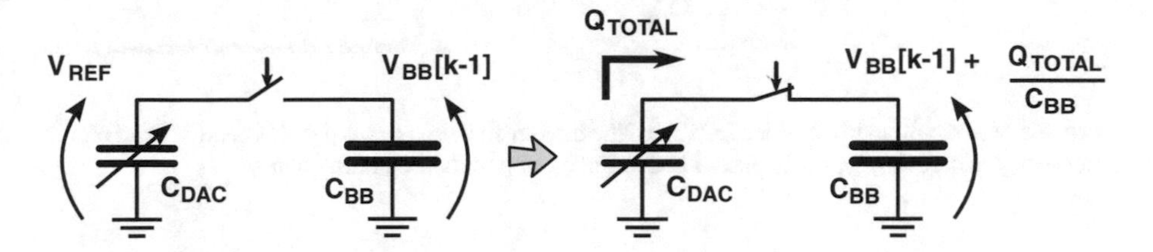

$$\begin{aligned} &C_{DAC_I+}[k] \cdot V_{REF_I+}[k] + C_{BB} \cdot V_{BB_I+}[k-1] \\ &\quad = (C_{DAC_I+}[k] + C_{BB}) \cdot \left(V_{BB_I+}[k-1] + \frac{Q_{TOTAL_I+}[k]}{C_{BB}} \right) \end{aligned} \tag{3.7}$$

Therefore, the instantaneous DAC capacitance $C_{DAC}[k]$ required to convey the wanted charge Q_{TOTAL} at time k is:

$$C_{DAC_I+}[k] = \frac{C_{BB}}{\dfrac{C_{BB}}{Q_{TOTAL_I+}[k]} \cdot (V_{REF_I+}[k] - V_{BB_I+}[k-1]) - 1}, \tag{3.8}$$

which is calculated for every new baseband sample, thus at F_S rate.

Again, accuracy can be improved by changing $V_{BB_I+}[k-1]$ at Eq. (3.8) to the average value $(V_{BB_I+}[k-1] + V_{BB_I+}[k])/2$, so that the trajectory of the baseband voltage between two samples is accounted in the C_{DAC} calculation.

3.2.3 Noise and Alias Performance

An important disadvantage concerning typical RFDACs is the fact that multiple noise contributors and spurs, including thermal noise, quantization noise and sampling aliases, are often upconverted to RF frequencies without any filtering. In this Section, the noise performance of the CQDAC is studied, in light of the previously discussed noise filtering capabilities given by charge-based architecture (Sect. 2.2.1).

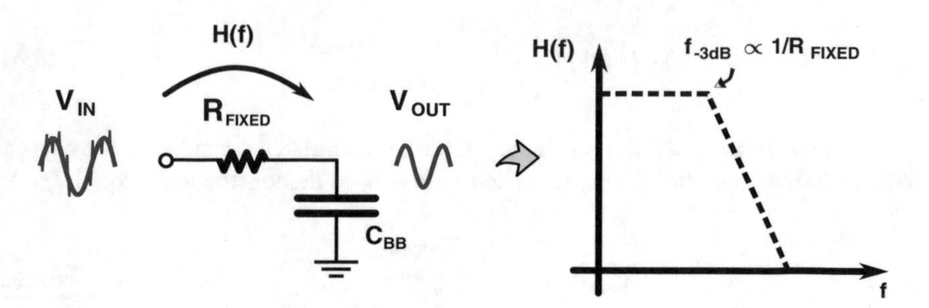

Fig. 3.6 Conventional first-order RC filter. The fixed resistance (R_{FIXED}) yields an also fixed cutoff frequency [Eq. (3.9)], typically placed above the maximum baseband frequency

3.2.3.1 Intrinsic R_{SC}C Noise Filtering

On analog-intensive transmitter architectures massive reconstruction filtering is typically applied in order to attenuate all sorts of noise contributors coupled to the baseband signal, including thermal and quantization noise. Regardless of the filter implementation, except for specific operating-mode reconfigurability [Sow09, Ros13, Ing13] the reconstruction filter cutoff frequency is always fixed and does not depend on the filtered signal itself. A first-order RC reconstruction filter is exemplified in Fig. 3.6, with its corresponding cutoff frequency given by Eq. (3.9).

$$f_{-3dB_{RC}} = \frac{1}{2\pi \cdot R_{FIXED} \cdot C_{BB}} \tag{3.9}$$

In the incremental-charge-based TX however, the two-phase operation of the QDAC switches leads to a Switched-Capacitor Resistance (R_{SC}) that, in combination with the baseband capacitor C_{BB}, produces an inherent single-order low pass filter in the signal path with a time constant that varies significantly over time. Since the DAC capacitance is a function of the input signal, it produces a time-varying $R_{SC}C$ filter (Fig. 3.7) whose bandwidth is automatically adjusted to accommodate the required instantaneous transitions of the baseband voltage, so that V_{BB} is not attenuated. In fact, with sufficient DAC resolution, it is possible to create a perfect distortion-free sinewave from a square-wave reference voltage toggling between V_{DD} and G_{ND}.

The noise sources which are not correlated to the input signal, on the other hand, are filtered by the charge-based architecture with an equivalent cutoff frequency given by the average conductance of CQDAC, as shown below:

$$f_{-3dB} = \frac{F_{LO} \cdot (C_{DAC_{RMS}})}{2\pi \cdot C_{BB}} \tag{3.10}$$

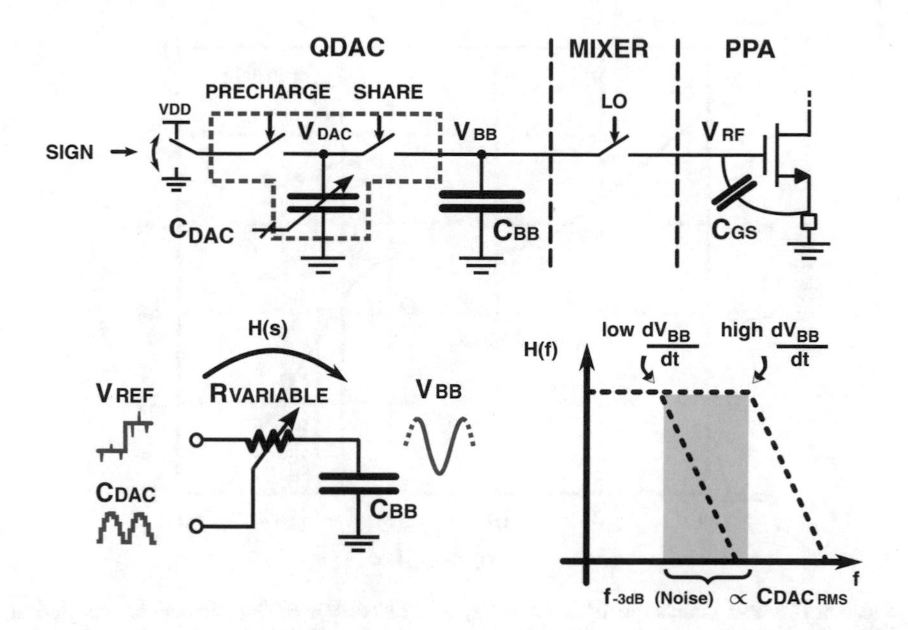

Fig. 3.7 CQDAC equivalent RC filter. The signal-dependent resistance ($R_{VARIABLE}$) creates a time-varying single-order low-pass filter whose bandwidth is automatically adjusted according to the input signal

As discussed in Sect. 2.2.1, the Root Mean Square (RMS) value of C_{DAC} is mostly a function of the signal's amplitude and frequency. As the signal swing decreases (decreasing the maximum signal derivative), a lower amount of DAC capacitance is needed to convey the required Q_{TOTAL}, decreasing the RMS value and consequently the noise cutoff frequency. An example periodic steady-state noise simulation (PNOISE) is shown in Fig. 3.8 for 8 MHz 400 mVpp single-tone. Even though the noise transfer function shows a cutoff frequency 2.8× lower than the fundamental, the signal itself is not attenuated.

The cutoff frequency variation with respect to the signal amplitude is shown in Fig. 3.9 using a 10 MHz single-tone. When Q_{TOTAL} is dominated by the baseband charge contribution, an increase in C_{BB} is followed by a directly proportional decrease in ($C_{DAC_{RMS}}$), keeping f_{-3dB} unaffected. In this case, for baseband voltage swings smaller than 80 % of the supply voltage (0.9 V in the example) the charge-based CQDAC provides an equivalent cutoff frequency that is notably smaller than the actual baseband frequency. This trend can also be seen in Fig. 3.10, where the result obtained from both PNOISE simulations and Eq. (3.10) are compared using baseband frequencies ranging from 1 to 20 MHz.

The same noise filtering capability is observed when wideband multi-tone transmit signals are applied. For example, the noise cutoff frequency of a 0.1 V_{RMS} 20 MHz BW WLAN-like modulated signal is roughly 4 MHz, corresponding to an RMS DAC capacitance of 1.1 pF.

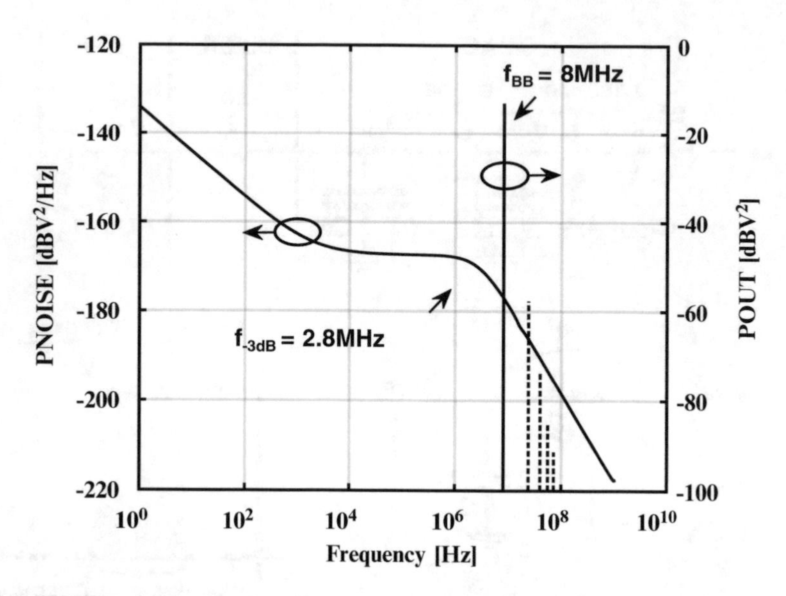

Fig. 3.8 PNOISE simulation of a 400 mVpp (−13.97 dBV), 8 MHz single-tone sampled at 128 MS/s. Uncorrelated noise is filtered by an equivalent cutoff frequency 2.8× lower than f_{BB}, without attenuating the baseband signal

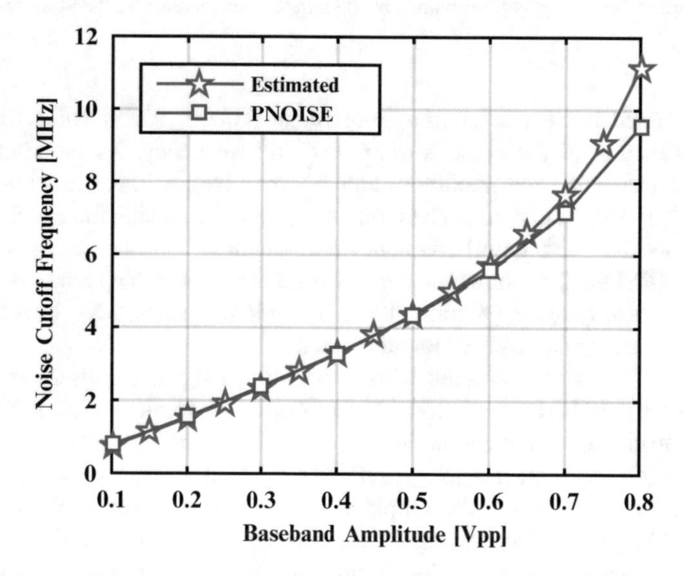

Fig. 3.9 Noise cutoff frequency versus baseband amplitude for a 10 MHz single-tone sampled at 128 MS/s, and a baseband capacitance of 50 pF

As demonstrated in Fig. 3.11, quantization noise is also filtered by this pole, relaxing the implementation and improving the power efficiency by allowing both sampling frequency and DAC number of bits to be reduced. Although a bit counter-

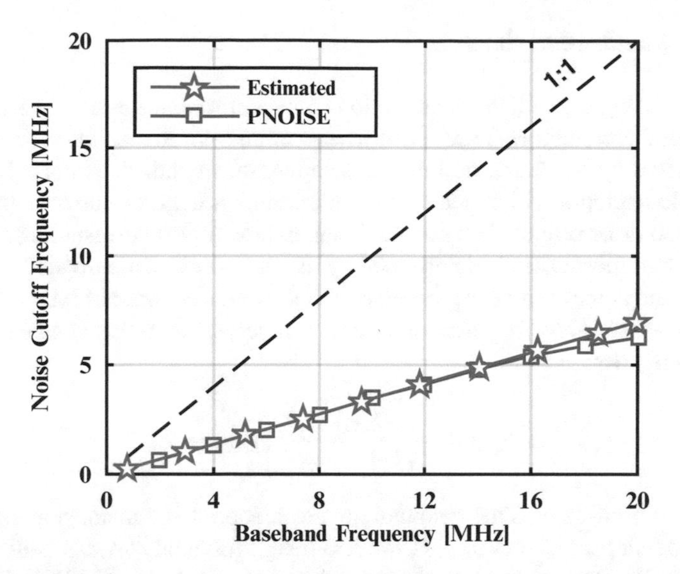

Fig. 3.10 Noise cutoff frequency versus f_{BB} for a 400 mVpp single-tone sampled at 128 MS/s, and a baseband capacitance of 50 pF

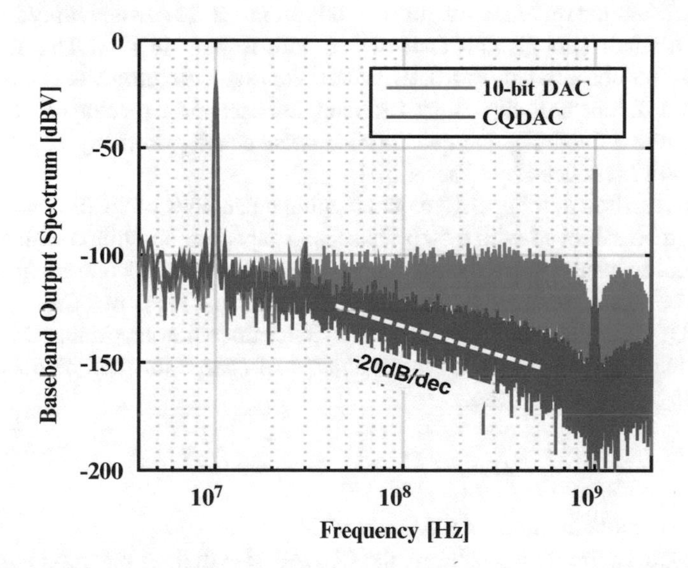

Fig. 3.11 Baseband output spectrum of the charge-based CQDAC versus a typical DAC implementation (C_{BB} = 50 pF, A_{BB} = 400 mVpp). The quantization noise filtering capability is clearly noted

intuitive when first considered, this remarkable effect of a time-varying cutoff frequency—that automatically scales with amplitude and frequency—is a major advantage of the incremental-charge-based transmitter.

3.2.3.2 Quantization Noise

Apart from being filtered, the quantization noise performance is also leveraged in the charge-based architecture. Introduced whenever an infinite resolution analog signal is represented by its discretized digital counterpart, quantization noise for a given DAC implementation is bounded by the minimum voltage or current step that can be produced at its output. In a conventional architecture it corresponds to the LSB size, which in most cases is determined by the full output scale divided by the total number of steps that can be represented with the given number of bits ($2^{\#BITS} - 1$). In a charge-based DAC however, the minimum voltage step that can be resolved at the output is given by:

$$\Delta V_{MIN} = \frac{C_{UNIT}}{C_{BB} + C_{UNIT}} \cdot (V_{REF} - V_{BB}) \tag{3.11}$$

Simple closed-form SNR calculations are hindered by an inevitable baseband signal dependence of ΔV_{MIN}. However, insightful analysis can still be made considering the design parameters involved: As shown in Eq. (3.11), the ΔV_{MIN} at the CQDAC output is fundamentally determined by the ratio between C_{UNIT} and C_{BB}. As a result, the quantization noise in this architecture can be reduced by simply choosing a small enough unit capacitance with respect to C_{BB}. This remarkable feature can be observed in Fig. 3.12, where the quantization noise floor is shown using two different unit capacitors for a single baseband capacitance of 50 pF. As expected, with a fixed C_{BB} the quantization noise density drops by 12 dB when the unit capacitor is reduced by a factor of 4.

Notably, as shown in Fig. 3.13 a quantization noise SNR of 86 dB (roughly 14-bit ENOB) can be achieved with a 50 pF baseband capacitor combined with a 2 fF unit capacitance. In cases where the unit capacitance cannot be decreased, quantization noise can be still be reduced by increasing the baseband capacitor C_{BB}.

The CQDAC total number of elements determines the maximum DAC capacitance, which translates into the largest amount of charge that can be transferred at once to the baseband capacitor.

$$Q_{TOTAL_{MAX}} \propto C_{DAC_{MAX}} = C_{UNIT} \cdot (2^N - 1) \tag{3.12}$$

where N is the total number of DAC elements.

As a result, rather than resolution the CQDAC size defines the maximum voltage step that can be produced at the output. Therefore, the charge-based DAC number of bits can be chosen based on a minimum required baseband frequency and voltage swing, as shown in Fig. 3.14 for multiple baseband amplitudes (considering a baseband capacitance of 50 pF). Lower bandwidth applications can therefore consume even less area, since a smaller number of elements is required for the same output swing.

3.2.3.3 sinc² Alias Attenuation

Finally, as discussed in Chap. 2 the baseband sampling alias attenuation should also be improved in the proposed architecture. In the capacitive charge-based TX, although the CDAC calculation is performed at baseband rate the charge transfer is transferred at LO frequency. Since the total charge required over one sampling period is subdivided in F_{LO}/F_S steps, a quasi-linear (*"L-fold"*) interpolation is inherently implemented between baseband samples (Fig. 3.15). Rather than a *sinc* transfer function, the sampling aliases in the charge-based transmitter are shaped by a *sinc*^{2} function, significantly increasing their attenuation.

The term "quasi" is used in this case due to the fact that the voltage step per LO cycle decreases within a single sampling period as C_{BB} charges (or discharges). This effect is specially noted when large F_{LO}/F_S ratios are used, without affecting however the alias attenuation, as demonstrated in Fig. 3.16. As noticed, a ratio of two between F_{LO} and F_S is already sufficient to improve significantly the alias suppression.

3.2.4 Harmonic Performance

On every transmitter architecture the spectral purity of the output signal is affected by each and every block composing the signal path. Current-source-based transmitters, for instance, have their harmonic performance affected by the inevitable

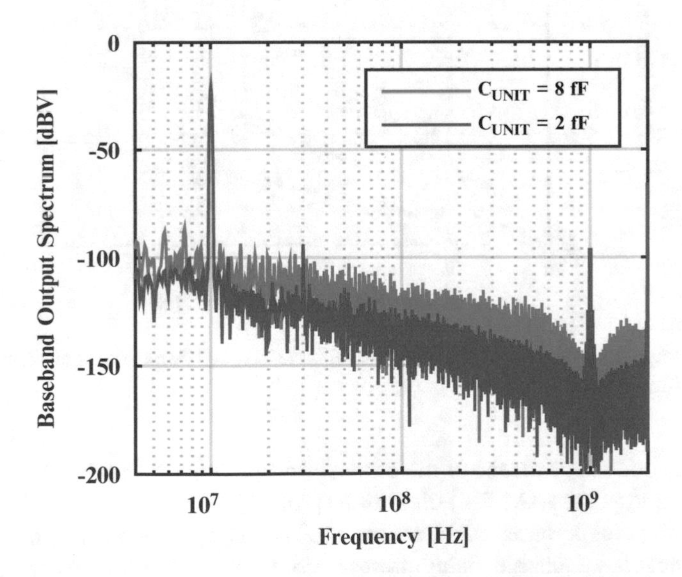

Fig. 3.12 Quantization noise floor versus C_{UNIT} for a fixed C_{BB} of 50 pF. As expected, the noise floor power density is reduced by 12 dB when the unit capacitance is divided by four

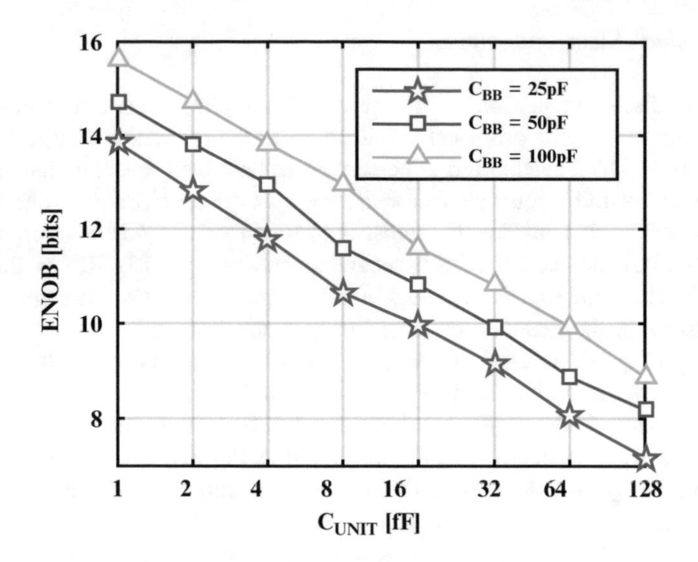

Fig. 3.13 CQDAC equivalent number of bits versus C_{UNIT}, for different baseband capacitances

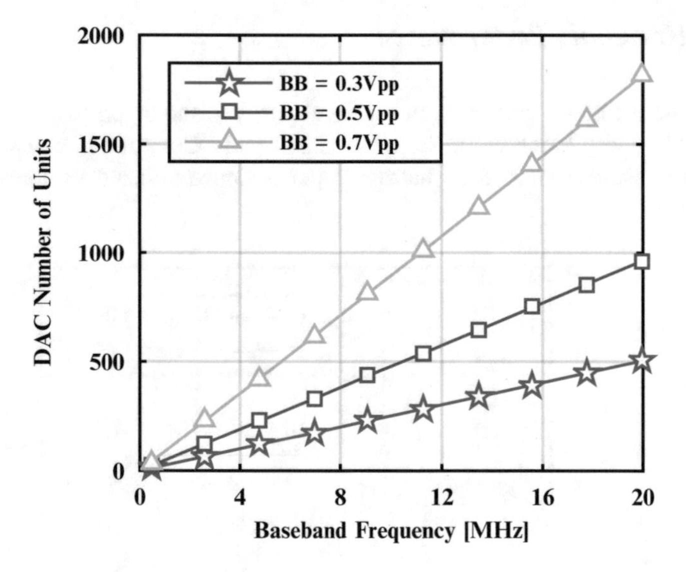

Fig. 3.14 Required number of DAC elements for various baseband maximum amplitude and frequency ($C_{BB} = 50\,\text{pF}$)

voltage and code dependence of the output current, typically modeled as amplitude (AM-AM) and phase (AM-PM) distortions [Ala14].

As for the charge-based architecture, the correct operation relies on precisely-defined packets of charge being transferred to the RF node every LO cycle. Due to incremental operation, errors in the charge transfer are also integrated at

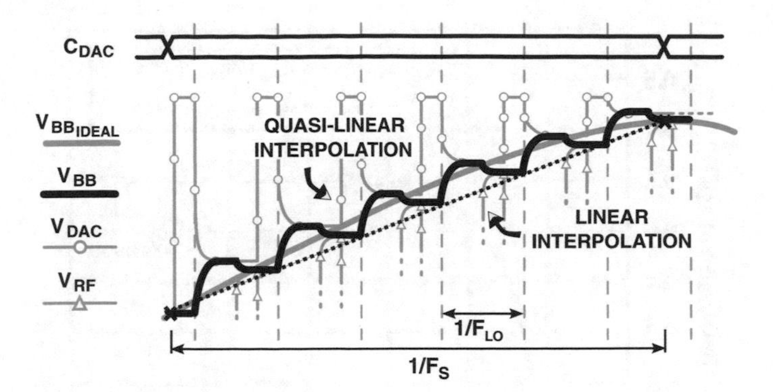

Fig. 3.15 Detailed timing diagram showing the "quasi" linear interpolation between samples achieved through the incremental charge of C_{BB}

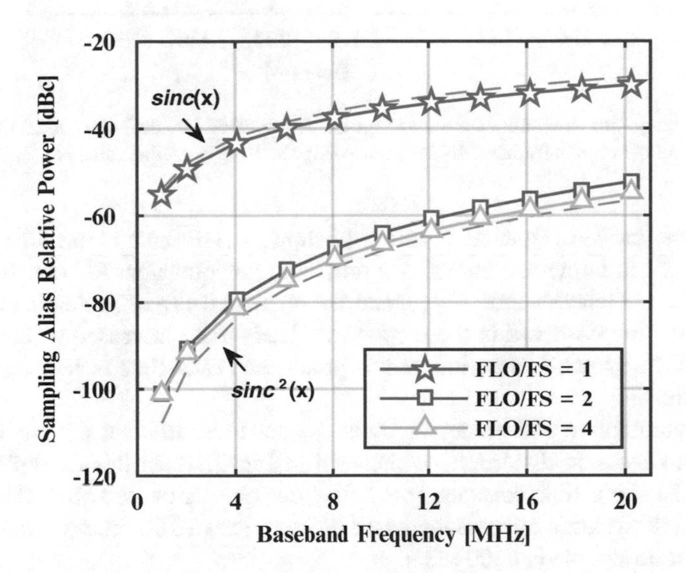

Fig. 3.16 Alias relative power versus baseband sampling frequency (f_S). For sampling frequencies lower than $f_{LO}/2$ the alias attenuation improves significantly

C_{BB}, potentially distorting the transmit signal and increasing the total harmonic distortion. Possible causes are either charge miscalculations or circuit non-idealities. However, while the former can be reduced after fabrication by calibration and fine adjustments of the charge calculation algorithm, the latter sets an intrinsic limit to the architecture's harmonic performance.

To achieve the required signal integrity, it is crucial to target complete voltage settling whenever charge is being conveyed along the signal path. As such, the switch ON-resistance (R_{SW}) is a major design parameter as it determines the path resistance and thus the time-constant of each phase in the charge transfer. An excessively large ON-resistance would hamper the required settling and degrade the

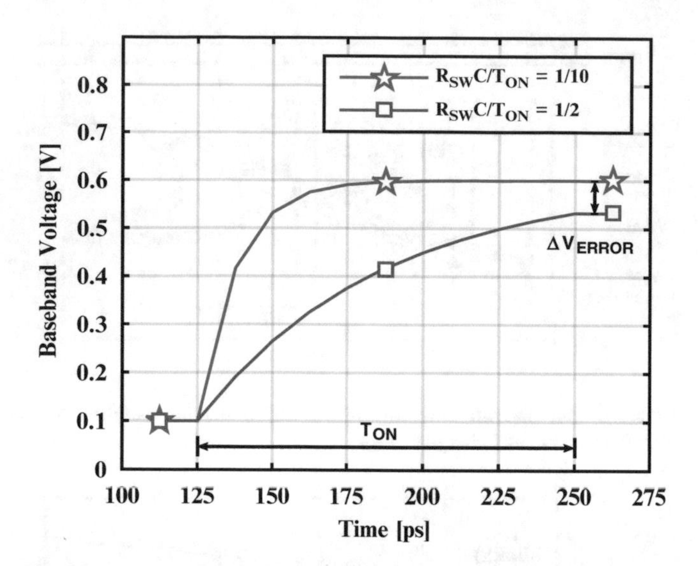

Fig. 3.17 Charge transfer and resulting voltage excursion using two different switch implementations. The voltage error introduced by the poor voltage settling is accumulated at C_{BB}

system dynamic performance. Figure 3.17 shows the difference in settling for two different switch implementations, one providing an equivalent RC constant of one tenth of the switch ON-time (T_{ON}), and the other half ($T_{ON}/2$). As demonstrated, the poor settling observed in the second case leads to an increased voltage error at the end of the switch ON period, that is propagated over time in the charge-based architecture.

The accumulation of the charge error at C_{BB} due to insufficient settling introduces significant harmonic distortion, as depicted in Fig. 3.18. In this example, just by reducing the $R_{SW}C$ time constant from half to one quarter of the switch ON-time the spurious-free dynamic range is increased by more than 10 dB, considering a 1 MHz baseband tone sampled at 200 MS/s.

On the other hand, increasing the switch aspect ratio (to reduce R_{SW}) also increases the power consumption of the LO drivers and introduces signal distortion due to a more pronounced switch charge injection. A good compromise can be found in Fig. 3.19, where the capacitive charge-based transmitter's harmonic performance is analysed against different settling conditions.

As expected, smaller time-constants (larger conductance) yield better harmonic performance. However, it can be also noted that for RC values bellow one seventh (1/7) of the switch ON-period, the harmonic distortion is not significantly affected by the switch conductance anymore, and HD3 settles around −71 dBc. This approach also allows one to define the minimum number of time-constants—and hence the maximum allowable switch resistance—corresponding to the maximum acceptable harmonic distortion, so as to optimize the switching power consumption in cases where less stringent linearity requirements apply.

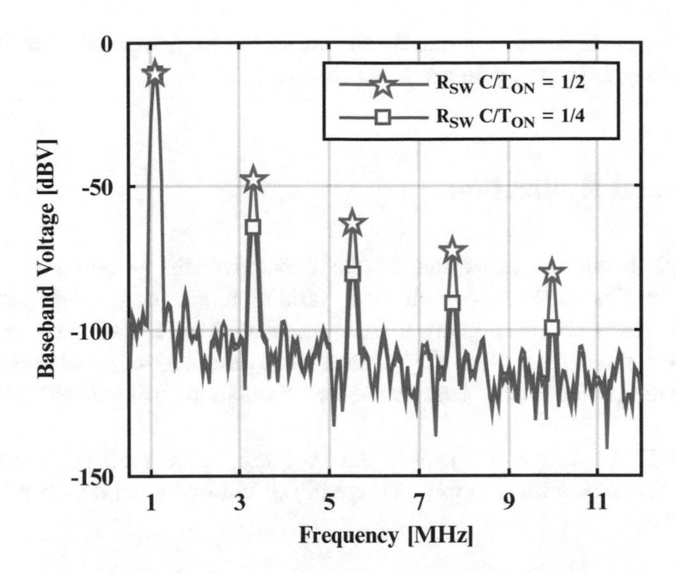

Fig. 3.18 Baseband output spectrum using two different settling conditions

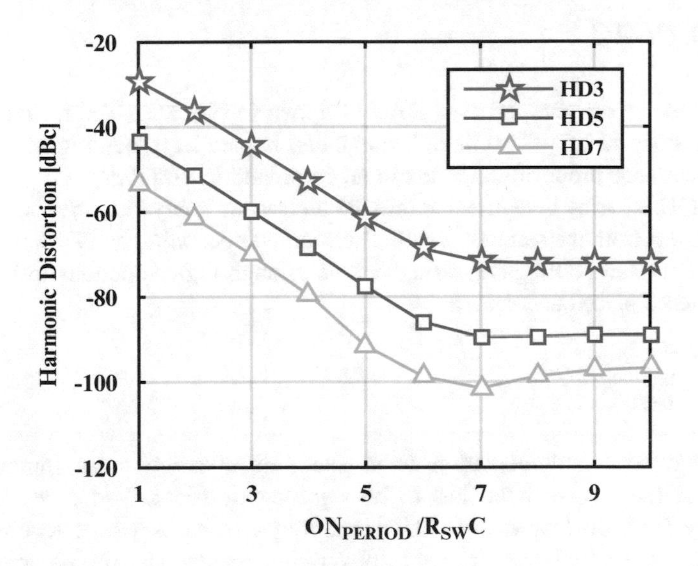

Fig. 3.19 Harmonic Distortion versus charge path time constant as a fraction of the switch ON-period

Charge injection is another potential source of non-linear distortion in charge-based architectures. Every time one of the switches in the signal path is closed, a certain amount of charge is absorbed by its inversion layer, which is again released when the switch is opened and the channel is extinguished. Even though the total amount of charge in the system is not altered, the charge balance between capacitances is distorted. Further details about the implementation of the several switches along the signal path are given in Sect. 3.3.

3.3 Circuit Realization

The complete circuit realization of the capacitive charge-based transmitter is disclosed in this Section. Architecture validation and early-stage performance evaluation were performed using a comprehensive MATLAB model that allowed system-level simulations observing minimum required out-of-band noise emission and harmonic performance. Transistor-level simulations using SPECTRE are also provided.

The CQDAC transmitter design and prototype were realized using a 28 nm CMOS technology, with a supply voltage of 0.9 V. Key layout details are provided in Sect. 3.3.1.4.

3.3.1 CQDAC

An overview of the charge-based DAC is shown in Fig. 3.20. The realized implementation consists of 1024 unit cells combined in parallel, providing an increasing DAC capacitance proportional to the input command *INPUT[0:9]*.

The CQDAC array is segmented into 5/5 binary and unary bits, with both of them being derived from the same unit cell. Thermometer decoding is fully implemented using NAND and NOR gates, using simple enough logic functions to have their layout custom placed and routed.

3.3.1.1 C_{UNIT}/C_{BB}

When SAW-less implementation is targeted, quantization noise (among other sources) at the receive band has to be significantly reduced in order to enable Frequency-Division Duplexing (FDD) operation. In cases where reconstruction filtering is not desired, the stringent noise requirements can only be achieved by either increasing the DAC number of bits, or the sampling frequency. In both cases power and area consumption are impacted.

Assuming a target quantization noise density of −165 dBc/Hz, it is required that:

$$10\log(\frac{e_{qns}^2}{A_{RMS}^2 \cdot (f_S/2)}) \leq -165 \quad (3.13)$$

where e_{qns}^2 and A_{RMS} stand for the RMS quantization error power and maximum signal amplitude. For a conventional DAC, e_{qns}^2 and A_{RMS} are both functions of the quantization step q_S as shown in Eq. (3.14):

$$e_{qns}^2 = \frac{q_S^2}{12}, A_{RMS} = \frac{2^N q_S}{2\sqrt{2}} \tag{3.14}$$

where N is the wanted DAC number of bits.

By further manipulating Eq. (3.13), it can be derived that:

$$N = \frac{\log_2\left(\frac{4 \cdot 10^{16.5}}{3f_S}\right)}{2} \tag{3.15}$$

indicating that more than 13-bit resolution is required if a sampling frequency of 500 MS/s or less were to be used.

On the capacitive charge-based architecture, on the other hand, the power and area consumption can be relaxed given that quantization noise is proportional to the ratio between two capacitances, namely the baseband C_{BB} and the QDAC unit capacitance (C_{UNIT}). As a result, the same 13-bit equivalent resolution can be achieved with several combinations of C_{UNIT} and C_{BB} (Fig. 3.21), allowing significant area reduction when both capacitances are decreased.

Besides quantization noise scaling, the baseband capacitance also has the important role of defining the noise cutoff frequency. As previously discussed, when low impedance RF loads are used, the equivalent noise $f_{-3\,dB}$ is shifted to higher frequencies, due to the increased conductance observed (through the passive mixer)

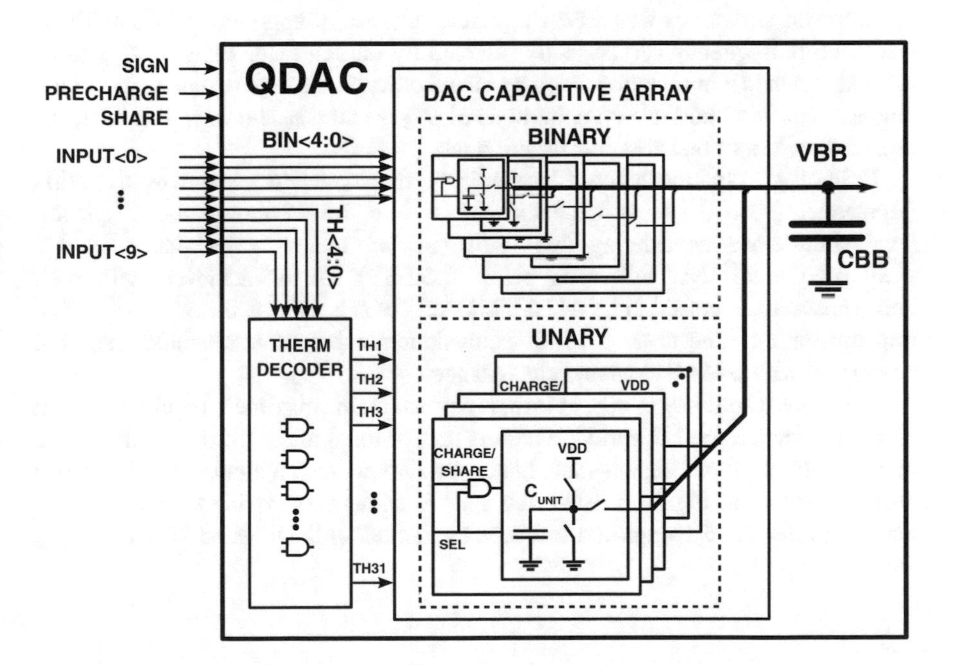

Fig. 3.20 Top-level schematic of the CQDAC implementation

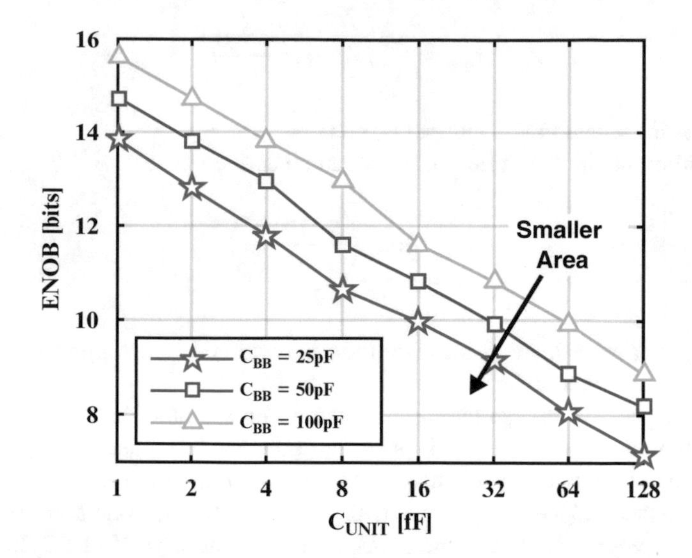

Fig. 3.21 CQDAC equivalent number of bits for various combinations of C_{UNIT} and C_{BB}

at the baseband side. If in-band noise filtering were to be provided in this case, a minimum C_{BB} value would be required in order to reduce the noise cutoff frequency sufficiently.

However, with an expected PPA input capacitance of approximately 250 fF, the noise cutoff frequency is marginally affected by the RF load, allowing C_{BB} to be reduced down to a minimum defined by quantization noise requirements. Therefore, among the several possible combinations of C_{UNIT} and C_{BB} shown in Fig. 3.21, the one leading to minimal area consumption was preferred.

Technology documentation files indicate that a relative standard deviation (sigma/mean) of 0.7 % is expected for a 2 fF MOM capacitance. System-level simulations considering this statistical information point that a 95.9 % yield (maximum INL and DNL below 0.5·LSB) can be achieved with such unit capacitance considering a 10-bit DAC, which is satisfactory for a first implementation. The desired 13-bit equivalent number of bits should therefore be assured with a 45 pF baseband capacitance.

The CQDAC number of bits, in turn, is defined by the maximum required amount of charge needed per LO period, which is proportional to the maximum derivative of the wanted output signal—and hence its maximum amplitude and frequency product. Shown in Fig. 3.22, 1024 unit cells is enough to provide a wide range of maximum baseband swings and frequencies, including the targeted 20 MHz $f_{BB_{MAX}}$ with 0.5 Vpp.

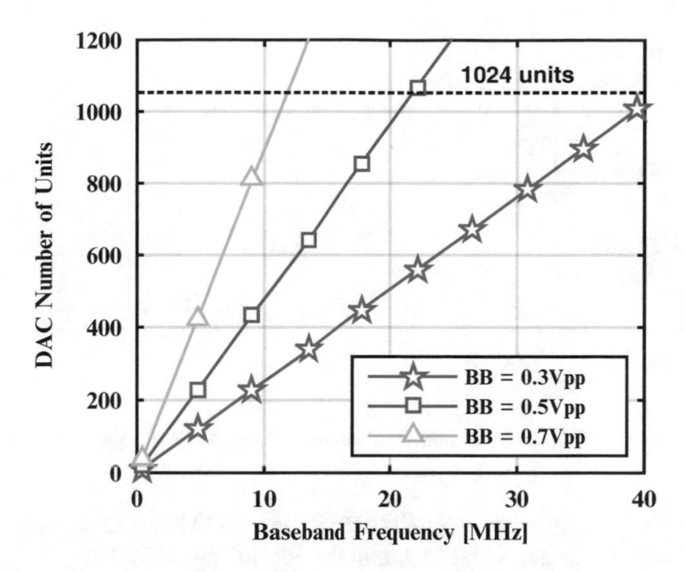

Fig. 3.22 Different possible combinations of maximum swing and frequency attained with 1024 DAC unit capacitances

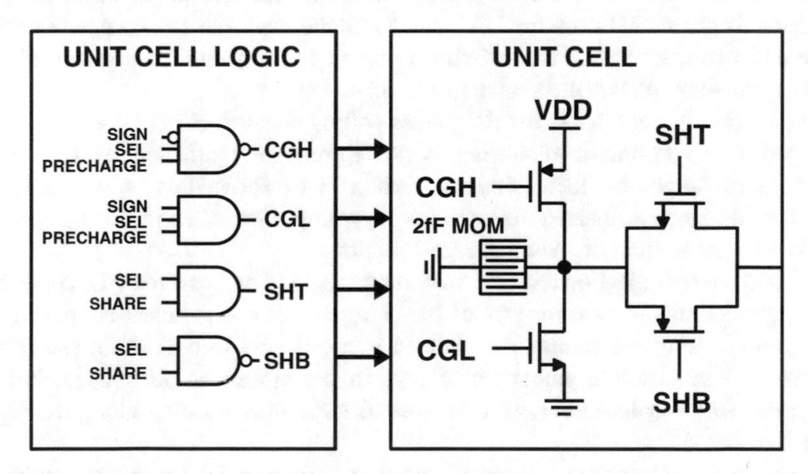

Fig. 3.23 CQDAC unit cell schematic, showing the control logic used to synchronize the different charge phases

3.3.1.2 Unit Cell

The CQDAC unit cell (Fig. 3.23) is a combination of a 2 fF Metal-Oxide-Metal (MOM) C_{UNIT} and four switch transistors, two to pre-charge the unit capacitor and two to provide a sharing path to the baseband capacitor.

Four signals control the unit cell operation: CGH, CGL, SHT and SHB. The *charge* and *share* control phases are respectively synchronized with non-overlapping

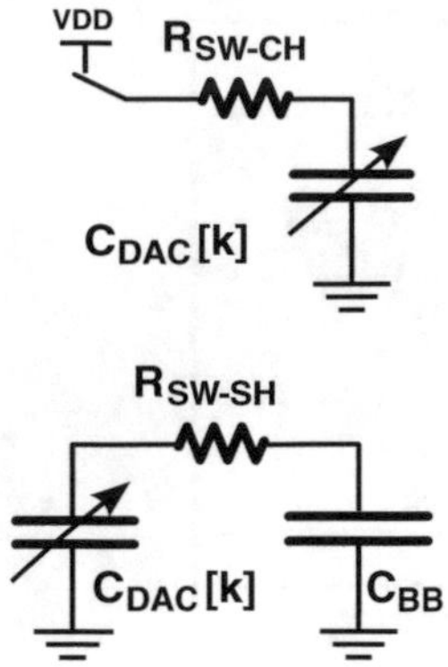

$$\tau_{CH} = R_{SW-CH} \cdot C_{DAC}[k] \tag{3.16}$$

$$\tau_{SH} = R_{SW-SH} \cdot \frac{C_{DAC}[k] \cdot C_{BB}}{C_{DAC}[k] + C_{BB}} \tag{3.17}$$

Fig. 3.24 Equivalent RC time constants involved in each one of the charge convey steps

25 % and 50 % duty-cycle PRECHARGE and SHARE LO signals, logically combined with the selection bit SEL and the control signal SIGN.

Again, complete voltage settling is crucial to achieve an improved harmonic performance. Whenever the DAC capacitance is pre-charged to one of the supplies or charge is shared between the DAC and baseband capacitances, complete settling should be observed before the switch is re-opened. The path time-constant of both *charge* and *share* phases are shown in Fig. 3.24.

As studied in Sect. 3.2.4, for RC values bellow one-seventh $(1/7)$ of the switch ON-period the harmonic distortion is not significantly affected by the switch conductance anymore. Increasing the switch sizes above this point would not improve the system linearity. On the contrary, larger switches in this case would imply a larger amount of switch charge-injection.

Switch charge-injection is a very important aspect to be observed when operating in charge-domain. Whenever one of the *charge* or *share* switches is opened, the charge populating the conduction channel is re-injected to both drain and source terminals. The absolute amount of charge in the system is not altered, but the expected charge balance between the several capacitors existing along the signal path is changed.

Considering the *share* phase as an example, the distortion mechanism works as follows (Fig. 3.25): when the SHARE switch is closed, the corresponding amount of charge required to build the inversion layer in the switch channel ($Q_{CHANNEL}$) is subtracted from both C_{DAC} and C_{BB}.

After the channel is formed, charge is shared between $C_{DAC}[k]$ and C_{BB} until their voltages equalize at V_{CLOSED}. Notice that in the V_{CLOSED} expression the channel charge is subtracted from the charge balance.

$$V_{CLOSED} = \frac{V_{DAC_N} \cdot C_{DAC}[k] + V_{BB_N} \cdot C_{BB} - Q_{CHANNEL}}{C_{DAC}[k] + C_{BB}} \tag{3.18}$$

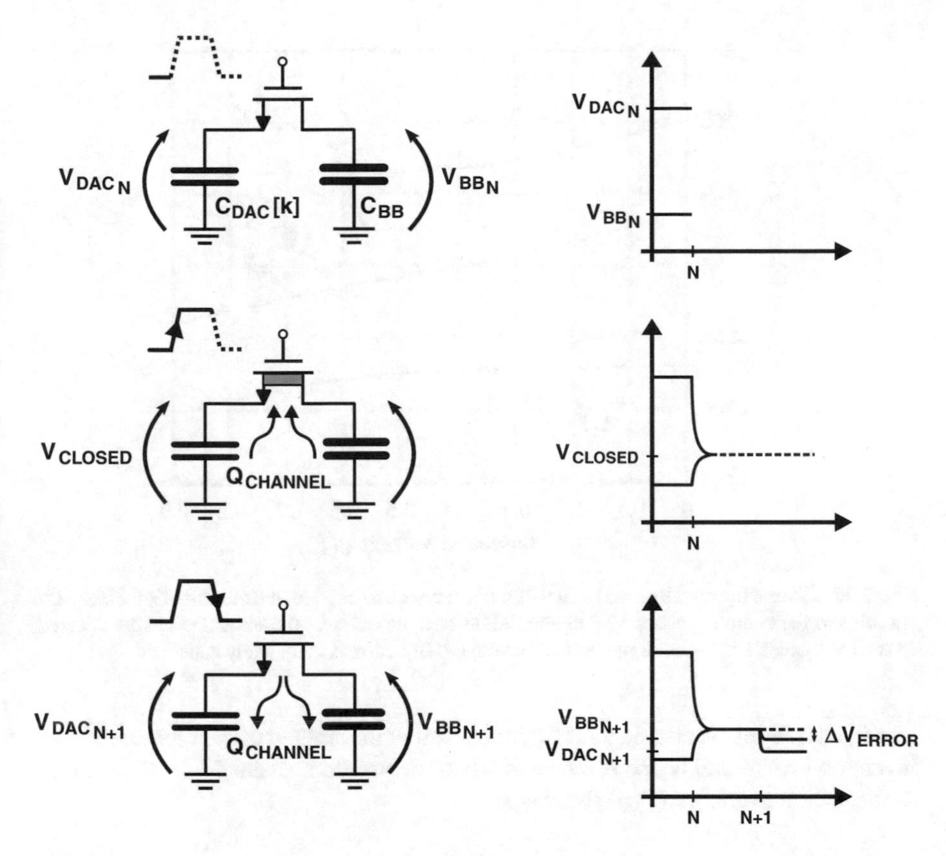

Fig. 3.25 Error voltage mechanism induced by charge injection

After complete settling, the switch is opened and the channel charge is again equally re-distributed, finally leading to the following baseband voltage:

$$V_{BB_{N+1}} = V_{CLOSED} + \frac{Q_{CHANNEL}}{2 \cdot C_{BB}} \tag{3.19}$$

Further manipulated to Eq. (3.20), it can be seen that the switch charge-injection introduces a voltage error (ΔV_{ERROR}) in the charge operation that is both proportional to $Q_{CHANNEL}$ and the code-dependent difference between the DAC and baseband capacitances.

$$V_{BB_{N+1}} = \frac{V_{DAC_N} \cdot C_{DAC}[k] + V_{BB_N} \cdot C_{BB}}{C_{DAC}[k] + C_{BB}} - \overbrace{\frac{Q_{CHANNEL}}{2} \cdot \frac{(C_{DAC}[k] - C_{BB})}{(C_{DAC}[k] + C_{BB})}}^{\Delta V_{ERROR}} \tag{3.20}$$

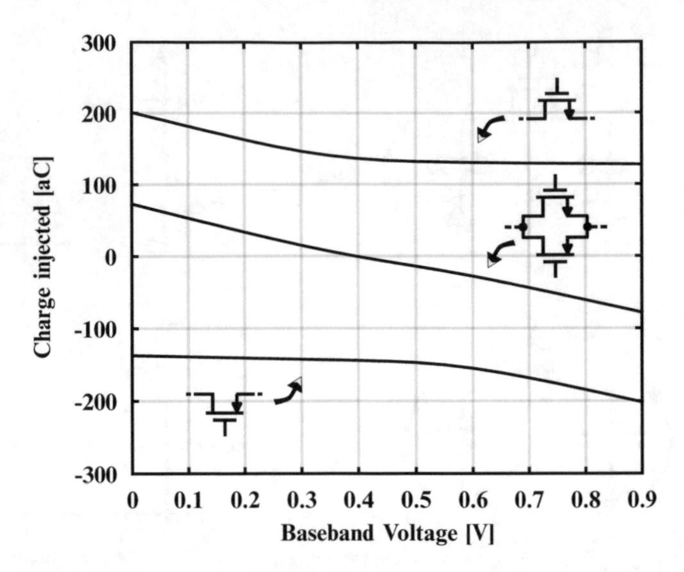

Fig. 3.26 Charge injected by a 400 nm/50 nm transistor as a function of its terminal's voltage. The combination of complementary NMOS and PMOS switches reduces the amount of charge injected, as well as creates a zero-crossing point where charge-injection is completely cancelled

Signal integrity is also degraded by the fact that the amount of charge stored in the inversion layer—and hence re-injected when the switch is opened—is proportional to the switch overdrive (V_{OV}) [Raz00]:

$$Q_{CHANNEL} = W \cdot L \cdot C_{OX} \cdot V_{OV} \tag{3.21}$$

In the case of the CHARGE switches, V_{OV} is always fixed since the source terminal is connected to a fixed potential. However, since both drain and source terminals of the SHARE switch follow the baseband voltage, signal distortion due to charge-injection voltage dependence is also introduced.

The best approach to mitigate the impact of charge-injection is by optimizing the switch sizes in order to reduce $Q_{CHANNEL}$, and applying complementary switches whenever applicable. As seen in Fig. 3.26, the complementary channel characteristics of NMOS and PMOS transistors reduces the amount of charge injected, which can be ultimately nulled at $V_{DD}/2$ by design. For a small price in design complexity and power and area consumption, the benefits of using complementary switches also include the reduction of clock feedthrough and increase in switch conductance, allowing the output signal range to extend from rail to rail.

The *charge* switches are kept either NMOS or PMOS to minimize the (non-linear voltage dependent) parasitic loading of C_{UNIT} given by their intrinsic capacitances. Finally, the schematic of the sized unit cell is shown in Fig. 3.27. As previously stated, all switches were sized in order to provide the required conductance corresponding to $1/7$ of the minimum switch ON period, given by 25 % of a 2.4 GHz

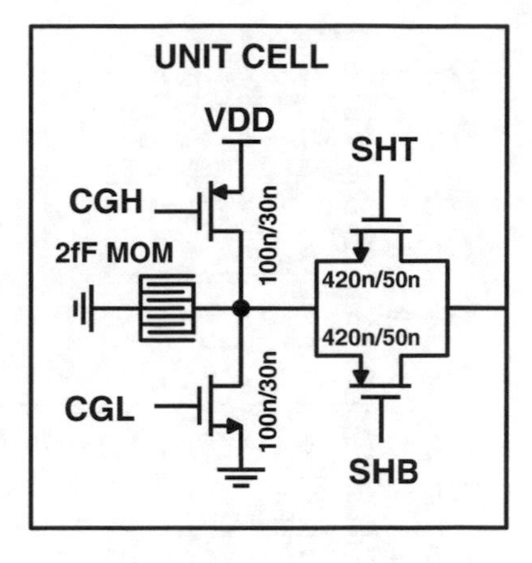

Fig. 3.27 Final schematic of the implemented unit cell showing device sizes

LO frequency. Minimum length is avoided at the *share* switches to reduce the OFF conductance and hence minimize leakage.

3.3.1.3 Thermometer Decoder

In a pure binary implementation, the unit cells composing the DAC architecture are binary-weighted and directly selected by the digital input word. Since no decoding logic is required, these architectures are simpler and typically more efficient. However, monotonicity is difficult to guarantee and large DNLs are thus typical. Unary implementations, on the other hand, are inherently monotone, but the decoding circuitry required to convert the binary input word into thermometer code can have a significant impact on both power and area consumptions.

A segmented implementation where the architecture is sub-divided into binary and unary cells can provide the best of both worlds. Depending on the number of unary cells, thermometer decoding can be implemented with simple circuitry.

In this implementation a 50 % segmentation was chosen, considering the still manageable thermometer decoder complexity. The least significant half of the input code is implemented using binary-sized unit cells, which are directly controlled by the digital input code. The most significant half, on the other hand, is implemented using 31 unary cells, all of them sizing 32 units. The corresponding unary part of the digital word is converted to thermometer code using the logic functions shown below [Van03].

$$TH_1 = \overline{\overline{\overline{B_0}.\overline{B_1}} + \overline{\overline{B_2}.\overline{B_3}}.\overline{B_4}}$$

$$TH_2 = \overline{\overline{\overline{B_2}.\overline{B_1}} + \overline{\overline{B_2}.\overline{B_3}}.\overline{B_4}}$$

$$TH_3 = \overline{\overline{\overline{B_0} + \overline{B_1}} + \overline{\overline{B_2}.\overline{B_3}}.\overline{B_4}}$$

$$TH_4 = \overline{(\overline{B_2}.\overline{B_3}).\overline{B_4}}$$

$$TH_5 = \overline{\overline{(\overline{\overline{B_0}.\overline{B_1}}.B_2) + B_3}.\overline{B_4}}$$

$$TH_6 = \overline{\overline{\overline{\overline{B_1} + \overline{B_2}} + B_3}.\overline{B_4}}$$

$$TH_7 = \overline{\overline{\overline{(\overline{B_0} + \overline{B_1})}.\overline{(\overline{B_1} + \overline{B_2})} + B_3}.\overline{B_4}}$$

$$TH_8 = \overline{\overline{B_3}.\overline{B_4}}$$

$$TH_9 = \overline{\overline{\overline{(\overline{B_0}.\overline{B_1})} + \overline{(\overline{B_1}.\overline{B_2})}.B_3}.\overline{B_4}}$$

$$TH_{10} = \overline{\overline{\overline{\overline{B_1}.\overline{B_2}}.B_3}.\overline{B_4}}$$

$$TH_{11} = \overline{\overline{\overline{(\overline{B_0} + \overline{B_1})}.\overline{B_3} + \overline{\overline{B_2} + \overline{B_3}}}.\overline{B_4}}$$

$$TH_{12} = \overline{(\overline{B_2} + \overline{B_3}).\overline{B_4}}$$

$$TH_{13} = \overline{\overline{\overline{\overline{B_0}.\overline{B_1}}.\overline{\overline{B_2}.\overline{B_3}}}.\overline{B_4}}$$

$$TH_{14} = \overline{\overline{\overline{\overline{B_1} + \overline{B_2}}.B_3}.\overline{B_4}}$$

$$TH_{15} = \overline{\overline{\overline{B_0} + \overline{\overline{B_1}.\overline{B_2}} + \overline{B_3}}.\overline{B_4}}$$

$$TH_{16} = B_4$$

$$TH_{17} = \overline{\overline{\overline{B_0}.\overline{B_1}} + \overline{\overline{B_2}.\overline{B_3}} + \overline{B_4}}$$

$$TH_{18} = \overline{\overline{\overline{B_2}.\overline{B_1}} + \overline{\overline{B_2}.\overline{B_3}} + \overline{B_4}}$$

$$TH_{19} = \overline{\overline{\overline{B_0} + \overline{B_1}} + \overline{\overline{B_2}.\overline{B_3}} + \overline{B_4}}$$

$$TH_{20} = \overline{(\overline{B_2}.\overline{B_3}) + \overline{B_4}}$$

$$TH_{21} = \overline{\overline{(\overline{\overline{B_0}.\overline{B_1}}.B_2) + B_3} + \overline{B_4}}$$

$$TH_{22} = \overline{\overline{\overline{\overline{B_1} + \overline{B_2}} + B_3} + \overline{B_4}}$$

$$TH_{23} = \overline{\overline{\overline{(\overline{B_0} + \overline{B_1})}.\overline{(\overline{B_1} + \overline{B_2})} + B_3} + \overline{B_4}}$$

$$TH_{24} = \overline{\overline{B_3} + \overline{B_4}}$$

$$TH_{25} = \overline{\overline{\overline{(\overline{B_0}.\overline{B_1})} + \overline{(\overline{B_1}.\overline{B_2})}.B_3} + \overline{B_4}}$$

$$TH_{26} = \overline{\overline{\overline{\overline{B_1}.\overline{B_2}}.B_3} + \overline{B_4}}$$

$$TH_{27} = \overline{\overline{\overline{(\overline{B_0} + \overline{B_1})}.\overline{B_3} + \overline{\overline{B_2} + \overline{B_3}}} + \overline{B_4}}$$

$$TH_{28} = \overline{(\overline{B_2} + \overline{B_3}) + \overline{B_4}}$$

$$TH_{29} = \overline{\overline{\overline{\overline{B_0}.\overline{B_1}}.\overline{\overline{B_2}.\overline{B_3}}} + \overline{B_4}}$$

$$TH_{30} = \overline{\overline{\overline{\overline{B_1} + \overline{B_2}}.B_3} + \overline{B_4}}$$

$$TH_{31} = \overline{\overline{\overline{B_0} + \overline{\overline{B_1}.\overline{B_2}} + \overline{B_3}} + \overline{B_4}}$$

These functions have the advantage of being fully implemented using NAND and NOR gates, and are simple enough to have their layout custom placed and routed.

3.3.1.4 Layout

An overview of the CQDAC floorplanning is given in Fig. 3.28. The DAC array is divided into 36 lines, with every line being occupied by a single binary or unary bit. The DAC lines are alternately placed above and below the central line, starting from the LSB (BIN0). In this way, the binary bits are gathered in the middle of the array, vertically surrounded by the unary cells. Control logic and thermometer decoder is placed on both sides.

The unit cells are uniformly spread along the bit line, so that horizontal gradients can be minimally accounted. Dummy cells are widely used to avoid border effects and to balance the capacitive loading seen from every control logic driver. The control signals are fed horizontally across the bit line (Fig. 3.29), and the unit MOM capacitance is shielded from the surroundings with top and bottom grounded metal plates.

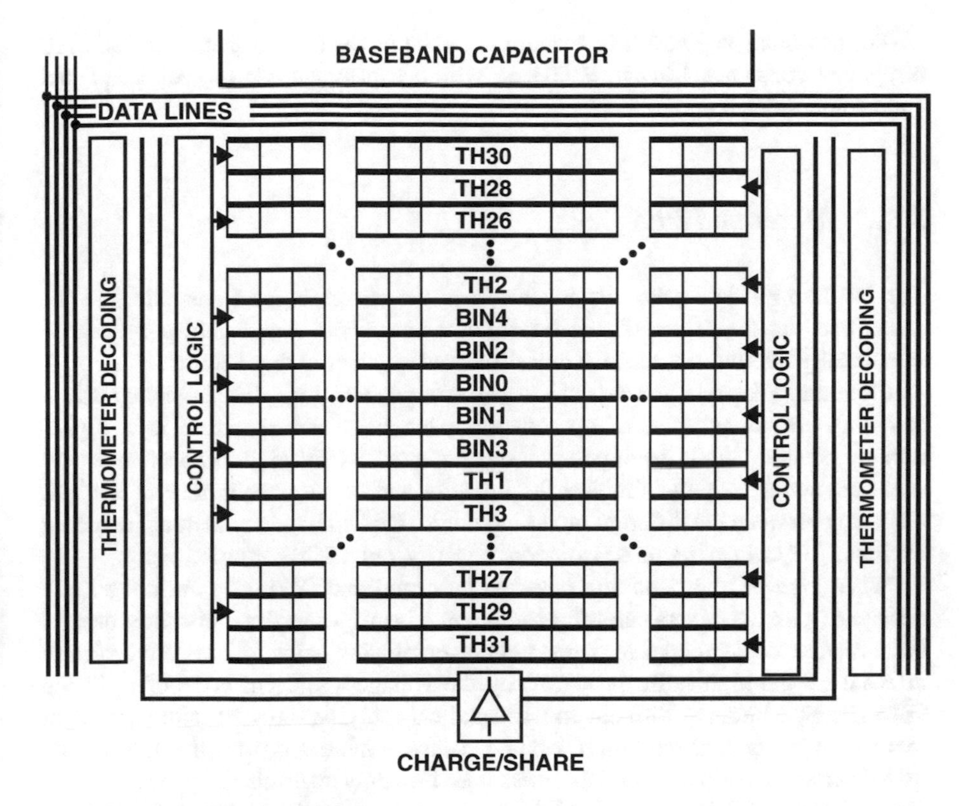

Fig. 3.28 CQDAC floorplanning

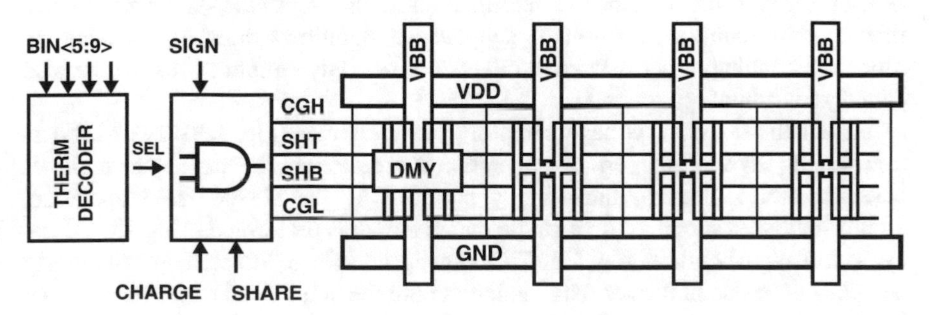

Fig. 3.29 Bit-line in detail

The *charge* and *share* signals are symmetrically distributed to both sides of the array. Phase noise constraints are relaxed here since phase information of both *charge* and *share* signals are "masked" by the corresponding mixer switch. Nevertheless, extensive buffering is still used as precaution to avoid that excessive delay creates overlap between the *share* and mixer phase.

The baseband 45 pF MOM capacitor is placed on the (north) side of the CQDAC array, and consumes 128 μm × 90 μm, with a capacitance density of 3.9 pF per square micrometer.

3.3.2 Mixer and PPA

The last two blocks in the transmit chain of the charge-based transmitter are the mixer and the PA driver. These two blocks have their design and performance intrinsically intertwined, and are thus discussed together in this Section.

From many aspects, the mixer is a key component of any TX implementation. It has the important role of up-converting the baseband spectrum to a higher output frequency, while driving the subsequent block's (PPA) input capacitance with sufficient swing, without limiting the linearity and noise performance of the TX. Choosing between the different mixer topologies depend on many factors, including available LO swings, required isolation, linearity, output flicker noise, etc.

When gain or high isolation between baseband and RF nodes are not strictly necessary, passive mixers can offer important advantages to the transmitter design. For instance, conventional implementations employing active mixers are typically affected by the high noise produced by the voltage-to-current conversion in the Gilbert cell, which are difficult to filter and can only be reduced with significant increase in power consumption [Oka11]. On passive mixers, on the other hand, since no V-I conversion is performed the noise floor is mainly determined by the switches' ON resistance. Flicker noise is also relaxed since there is no bias (DC) current flowing through the switches [Mir11a]. Finally, the transistor stacking of active mixers also require higher voltage supplies—a significant drawback considering modern technology nodes. Passive mixers are not only simple to realize but also friendly to technology scaling.

For all the above, a voltage sampling passive mixer (Fig. 3.30) is realized in this design [He09]. Voltage-mode operation is preferred in this case since it allows baseband noise filtering, intrinsically provided by the charge-based operation. Double-balanced mixing with a single-ended output is performed using 25 % duty-cycle LO signals, summing I and Q signals in voltage domain by alternately sampling each one of the baseband voltages onto the output load capacitance, given by the input capacitance of the subsequent PPA.

The PPA, in turn, is the last block in this charge-based TX implementation. It amplifies the upconverted RF signal from the mixer output and feeds it to an off-chip 50 Ω load representing the PA input. As with mixers, key PPA design requirements are low noise contribution for FDD operation and high linearity for improved EVM and ACPR performance. Power efficiency is another important concern, however to sustain sufficient linearity many times the use of less efficient amplifier topologies (e.g. Class-A) becomes necessary. In these cases, the PPA design becomes in practice a matter of finding the best spot with regard to linearity and noise performance, with little room left for improvements in power efficiency.

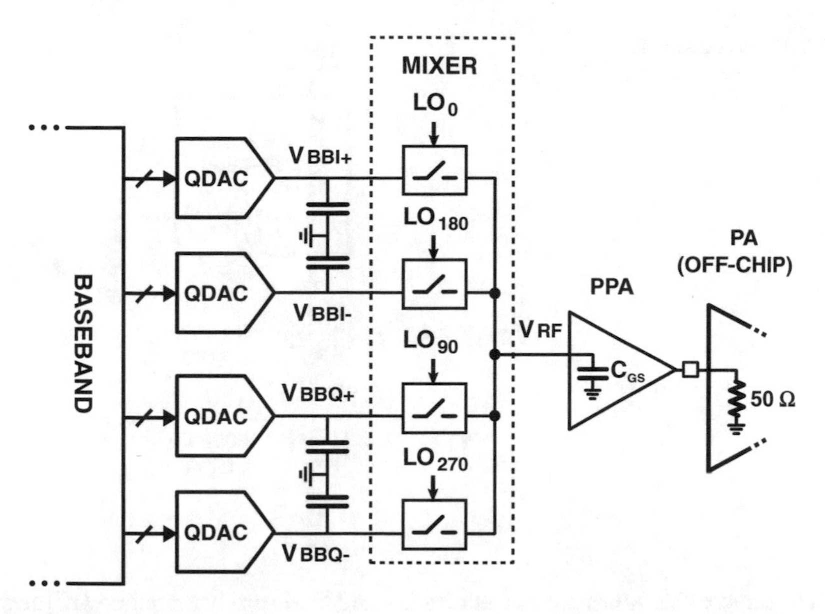

Fig. 3.30 CQDAC TX block diagram showing the voltage sampling passive mixer and PPA used in this implementation

If the input signal characteristics are fixed, one of the few things available to relax the design of a Class-A PPA is to increase the supply voltage (if possible). A larger supply voltage enables larger output swings that, for a constant output power: first, decreases the drain current and hence the required transistor width, incurring less capacitance to be driven by the preceding stage; second, improves the PPA linearity by providing more voltage headroom, and third, relaxes the design and decrease the losses in the output matching network [Raz12]. However, together with a larger supply voltage come reliability issues. In order to prevent transistor breakdown, the inclusion of a thick-oxide cascode transistor becomes necessary.

3.3.2.1 PPA Design

The PPA schematic is shown in Fig. 3.31. It consists of a Common Source (CS) amplifier, cascoded with a thick oxide device allowing 1.8 V supply. An external bias "tee" is used to increase the maximum output swing.

A compression point of 10 dBm (2 Vpp) is targeted in this implementation, demanding a minimum gain of 12 dB from the PPA, considering the pre-fixed maximum baseband swing of 0.5 Vpp.

Though not obligatory, the CQDAC output swing is maximized when the DC voltage is set to $V_{DD}/2$ (0.45 V). With a pre-defined overdrive, the input transistor M1 is sized to provide the required transconductance for the given gain and output load (50 Ω), also accounting for the current division between $1/gm_{M2}$ and gds_{M1}.

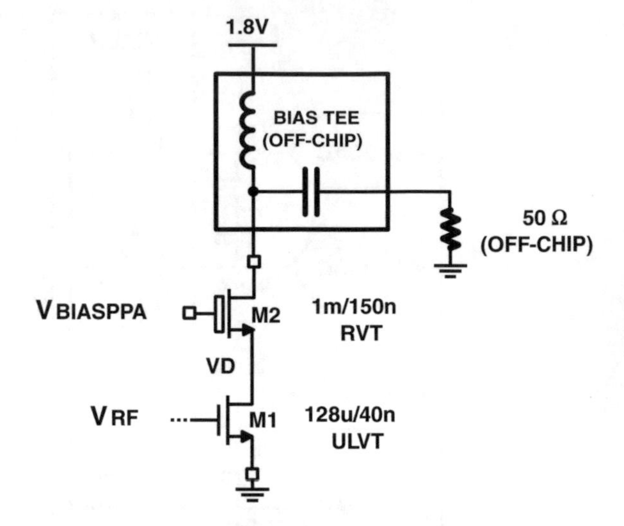

Fig. 3.31 PPA schematic depicting the external bias tee and 50 Ω load

The thick-oxide transistor M2 in turn is sized to ensure the required OP1dB. As in typical PPA designs, the cascode bias voltage ($V_{BIASPPA}$) sets a tradeoff between linearity and stress. Increasing the bias voltage can possibly drive M2 to triode, changing VD and thus creating compression. Decreasing $V_{BIASPPA}$, on the other hand, may produce excessive stress on M2 due to an increased drain-to-source voltage. In this design $V_{BIASPPA}$ is made equal to 1.3 V, and is provided externally from a dedicated pin so that tunability is improved.

To avoid gain degeneration due to excessive voltage drop at the bond-wires, three bond-pads were dedicated for ground connection. Figure 3.32 show the simulated compression characteristic and efficiency versus input power. Output noise spectral density is shown in Fig. 3.33.

After extraction, a total gate capacitance of approximately 250 fF including layout parasitics is expected.

3.3.2.2 Mixer Design

Simple in concept and design, the passive mixer used in this CQDAC transmitter realization consists of four switches that sequentially sample each one of the quadrature baseband voltages to the output node at the corresponding LO phase (Fig. 3.34).

When any of the mixer switches is closed, charge is shared between the corresponding baseband capacitance and the C_{GS-PPA}, instantaneously driving the PPA input voltage to the corresponding IQ component. Not to degrade the TX performance, the mixer switches should be designed accordingly observing minimum constraints in terms of signal integrity and noise contribution.

Similar to the CQDAC, the mixer switches are also part of the charge-based operation and thus must obey the maximum time constant requirement, set previously through system analysis. As shown in Eq. (3.22), the time constant given

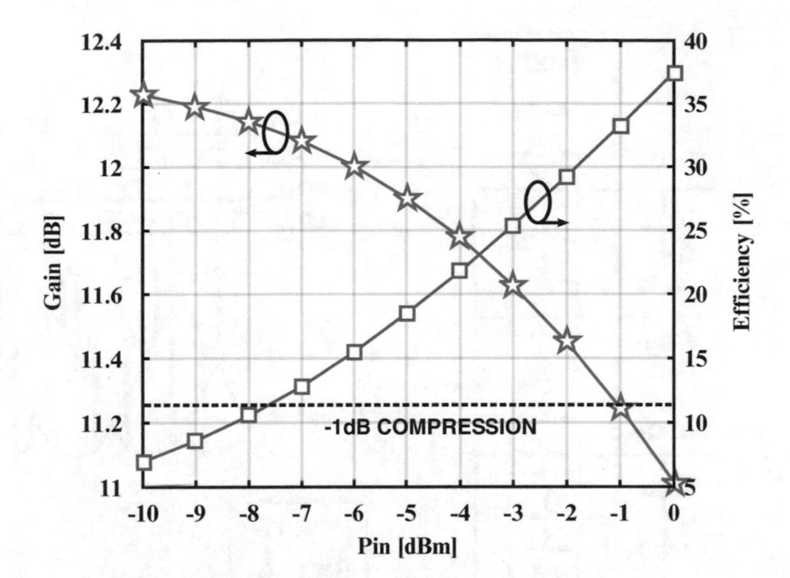

Fig. 3.32 Simulated PPA gain and efficiency

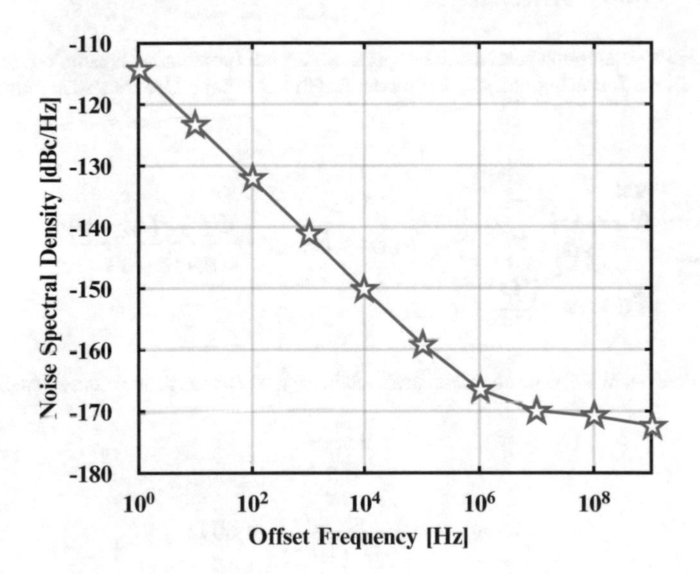

Fig. 3.33 PPA output noise spectral density

by the mixer switch (Fig. 3.35) in combination with the baseband and PPA input capacitances should be smaller than $1/7$ of 25 % of one LO period, requiring a maximum switch resistance of 71 Ω.

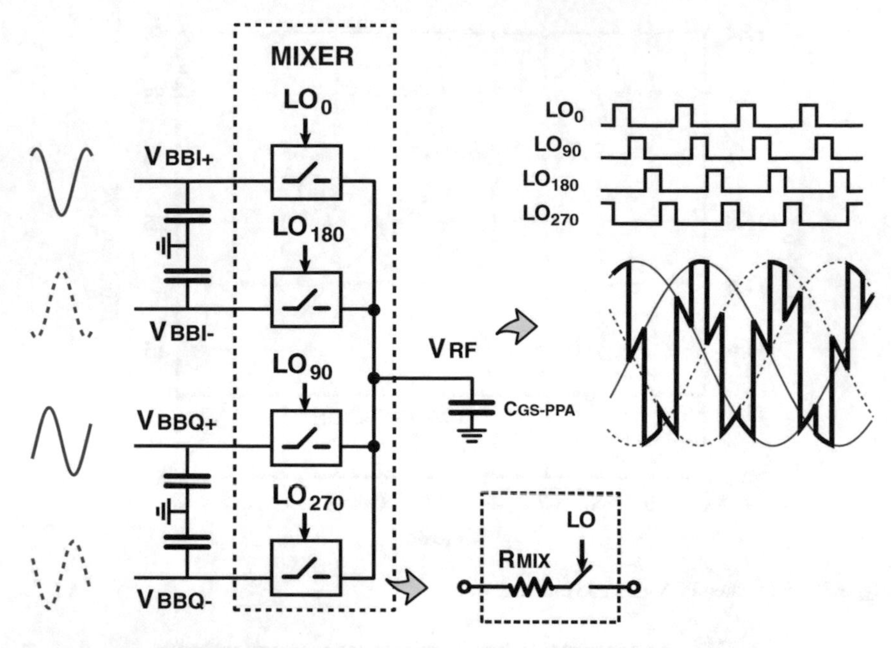

Fig. 3.34 Voltage sampling passive mixer operation. Each IQ component is sampled for 1/4 of the LO period. In the following analysis the mixer switch has a finite ON resistance represented by R_{MIX}

RMIX

CBB CGS-PPA

$$\tau_{LO} = R_{MIX} \cdot \frac{C_{BB} \cdot C_{GS-PPA}}{C_{BB} + C_{GS-PPA}} \tag{3.22}$$

Fig. 3.35 Equivalent RC time constants involved in each one of the charge convey steps

$$R_{MIX_{MAX}} = \frac{\left(\frac{0.5}{2\sqrt{2}}\right)^2}{4kT \cdot 10^{16.5}} \approx 60\,\Omega \tag{3.23}$$

With respect to noise, it can be proved that the noise contribution of all four switches in a 25 % duty-cycle passive mixer can be simplified as a single white-noise voltage source at the RF side with a power spectral density equal to $4kTR_{MIX}$, where R_{MIX} is the switch ON resistance [Mir11b]. If the duty cycle is decreased to something less than 25 %, the resulting "hold" phase shapes the noise spectrum by concentrating a larger part of the noise power at low frequencies [Kun06]. Nevertheless, for the given purposes the 25 % assumption is valid in cases where the

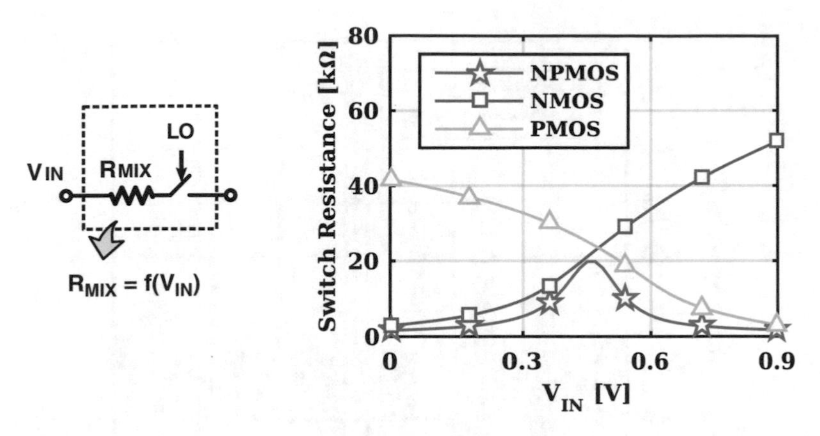

Fig. 3.36 Switch resistance voltage dependence for example NMOS, PMOS and NPMOS implementations

duty cycle is slightly decreased to prevent overlapping between the several switch phases.

To make sure that the mixer noise does not violate the maximum out-of-band noise emission, the switch ON resistance corresponding to a spectral noise density of −165 dBc/Hz is calculated, considering a 0.5 Vpp baseband swing:
Therefore, in this realization the minimum switch conductance requirement is overruled by noise constraints.

It is pointed in literature that transmitter implementations using advanced technology nodes should not have their linearity limited by the passive mixer, since switching performance is always improving over time. Instead, Pre-Power Amplifier nonlinearities would be the limiting factor in this case [Mir11a]. Though partially true, this statement does not take into account the considerable distortion introduced by the switch conductance voltage dependence given by a non-fixed switch overdrive, as shown in Fig. 3.36.

To illustrate the point, Fig. 3.37 demonstrate two spectra: one using an ideal switch with a finite ON resistance of 50 Ω, and another with an NPMOS switch providing a peak R_{MIX} of the same value. Even with a much reduced voltage dependence achieved with the complementary switch (with respect to a single NMOS or PMOS), the third-order harmonic is degraded by roughly 30 dB.

The approach used in this implementation to improve the mixer linearity was to increase the switch overdrive by decoupling the LO signal and biasing the transistor gates through a 15 kΩ bias resistor. By shifting in design the DC value by 0.2 V, not only the mixer linearity could be improved significantly, but the mixer switch sizes could also be reduced, incurring less charge-injection (already minimized through NPMOS switch utilization).

Finally, the mixer switch schematic can be seen in Fig. 3.38.

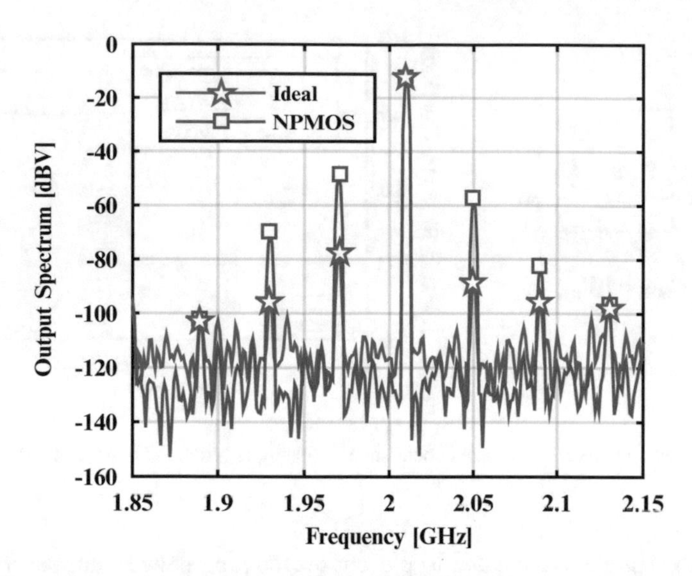

Fig. 3.37 Example spectrum showing the impact of the switch resistance voltage dependence

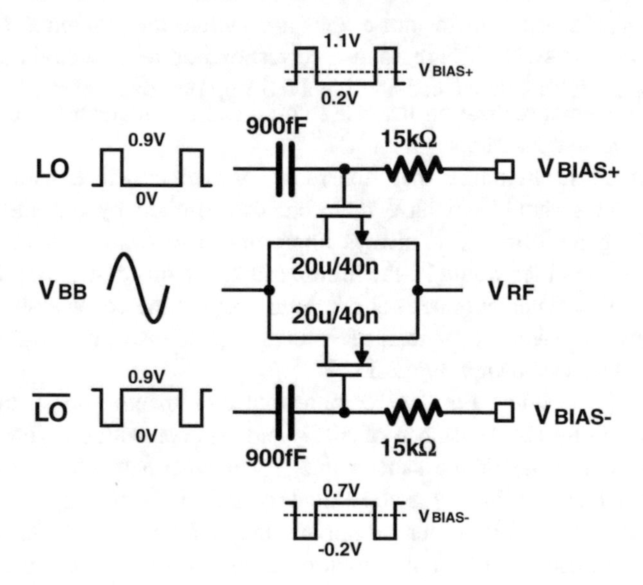

Fig. 3.38 Mixer switch schematic with device sizes in detail. The LO signal is decoupled and biased with V_{BIAS+} and V_{BIAS-} in order to increase the switch overdrive

3.3.3 LO Generation

The LO generation block is responsible for creating the four duty-cycled quadrature LO signals, which are used both to drive the voltage sampling passive mixer and to synchronize the different charge-based operations.

Two important aspects should be observed when designing the LO generation block: First, due to the charge-based nature of the complete transmitter it is crucial that no phase overlapping is observed, neither among the mixer switches nor the different *charge* and *share* phases operating the QDAC. While the latter would provide a short path between the supply and C_{BB}, the former causes undesired charge sharing between the different baseband capacitors. In both cases the incremental charge accumulation would be corrupted, leading to severe signal distortion that could hardly be compensated during CQDAC operation.

Second, LO phase-noise from the transmit path is fully translated to the RF output and therefore must be minimized. An excessive phase noise skirt can mask the signal received by a neighbor user in close proximity, or by its own receiver in full-duplex operation. For the latter, the same stringent phase noise requirement of −165 dBc/Hz at 45 MHz offset is targeted at the LO generation block output, considering the noise budget described in Chap. 1.

A top-level block diagram of the LO generation block is shown in Fig. 3.39. Starting from an external signal source at twice the LO frequency ($2 \cdot f_{LO}$), the input 2LO signal is first made differential using a hybrid coupler and AC coupled to the chip. On chip, a first amplifying stage (inverter) converts the sinusoidal input into a more "square wave"-like signal. The duty-cycle at the output is controlled by adjusting the DC voltage at the inverter input, done with replica biasing using tunable NMOS and PMOS strengths. With duty-cycles ranging from 46 % to 50 %, the intermediate LO signals at $2 \cdot f_{LO}$ are divided by two in frequency using D flip-flops, specially designed to provide the required speed with minimal phase noise constraints. The divided output is finally combined using logic gates with the signals $2LO+$ and $2LO-$, providing the four quadrature LO phases with duty-cycles that can be adjusted from 23 % to 25 %.

The only role of the divided output in this case is to provide a logical "mask" to which the $2LO+$ and $2LO-$ signals are combined. The phase noise of the output LO signals, as a result, are exclusively defined by the higher frequency $2 \cdot f_{LO}$ signals, allowing the noise constraints and thus power consumption of the frequency divider to be reduced significantly.

3.3.4 Top-Level Description

A top-level schematic of the implemented prototype is shown in Fig. 3.40. Different blocks such as a network-on-chip (NOC), clock dividers, memory and voltage buffers are also included.

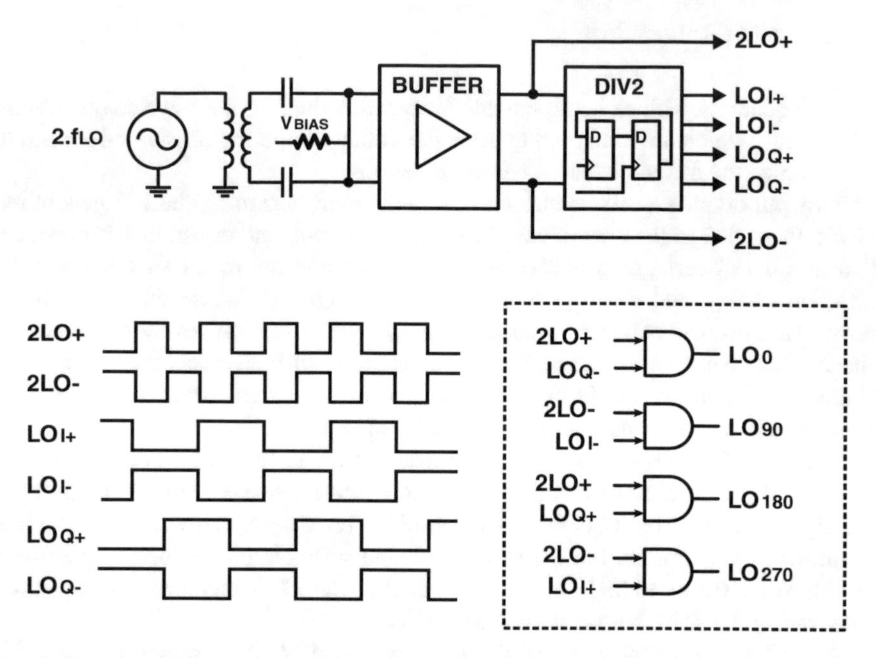

Fig. 3.39 LO generation scheme. Quadrature LO signals with adjustable duty-cycle are made from logic combination between LO signals at f_{LO} and $2 \cdot f_{LO}$ frequencies

For a given transmit signal, the charge calculations and the corresponding DAC capacitances are first evaluated externally using mathematical tool MATLAB, and later loaded into the integrated memory. Once loaded, the data is continuously cycled, first being synchronized to the appropriate LO phase and then fed to the corresponding CQDAC. The memory readout can be done at multiple fractions of the LO frequency (from 1/1 to 1/8), as set by a sequence of frequency dividers that can be bypassed if desired.

This first charge-based TX realization features a 1K latch-based integrated memory that can be programmed via Serial Peripheral Interface (SPI) interface. Each memory position is 24 bits wide, originally designed to provide a 12-bit resolution to both I and Q data. However, due to the fact that in the charge-based TX the DAC capacitance values fed to each one of the QDACs can be distinct at any given time ($D_{I+} \neq D_{I-}$), the number of bits per IQ differential component would have to be reduced to 6 in this case. Therefore, to keep the required resolution the IQ data is interleaved (D_{I+}/D_{I-}, D_{Q+}/D_{Q-}, D_{I+}/D_{I-}, ...) before being loaded to the memory, reducing by half the number of unique IQ samples that can be transmitted (from 1K to 512 words). This memory limitation prevented the EVM characterization of the first prototype since not enough symbols could be stored on chip. Nevertheless, having an integrated memory simplifies the design reducing the number of pads required and allowing the sampling frequency to be increased up to 2.4 GHz.

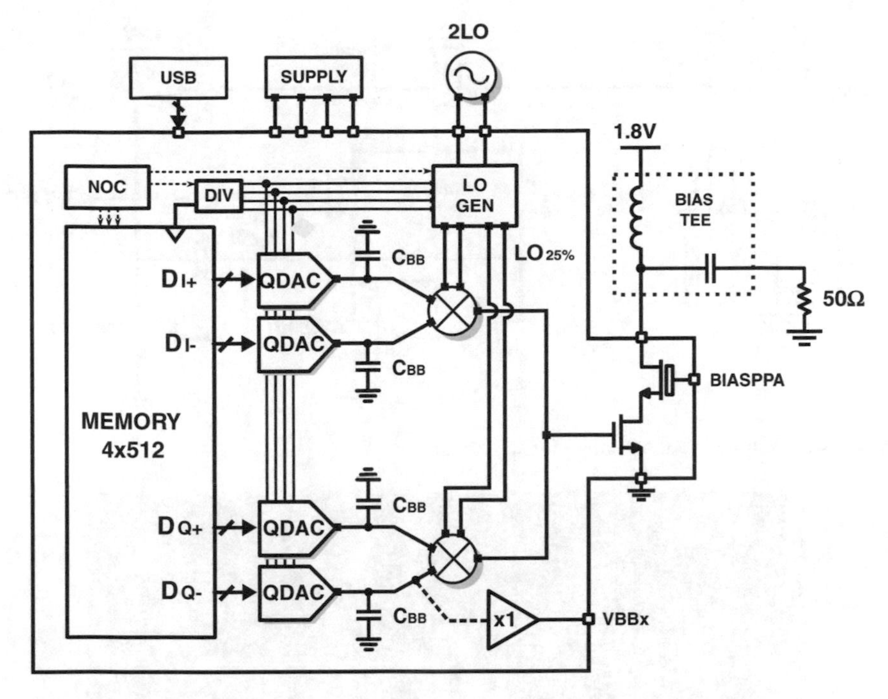

Fig. 3.40 Top-level block diagram of the CQDAC TX prototype

An additional test mode was implemented where the mixer block can be independently switched OFF and the baseband voltages are observed without being influenced by the RF load. By removing the RF charge component, valuable information about the CQDAC characteristics and intrinsic performance (including linearity) can be assessed. For this test, unity-gain amplifiers working as voltage buffers were included in order to probe the baseband voltages without adding external interference.

The voltage buffers were implemented as two-stage Miller-compensated OTAs in closed loop for unity-gain. Each amplifier provides a DC gain of 60 dB, with a Gain-Bandwidth Product (GBW) of 200 MHz. To make sure that observed harmonic performance is not affected by the amplifier, the buffer was implemented using 1.8 V thick-oxide transistors, with PMOS input pair. To guarantee stability at all conditions, a minimum of 55.6° phase margin is assured even when the output node is loaded with 10 pF capacitance. The schematic is shown in Fig. 3.41.

Finally, the control bits existing in every block are set through a dedicated Network-On-Chip (NOC) whose internal registers can be programmed via an USB interface. As early stated, the proposed CQDAC transmitter was prototyped using a 28 nm CMOS technology. The chip occupies $1.4 \times 1.1\,\text{mm}^2$ (with pads), with an active area (including the entire transmitter except integrated memory and NOC) of $0.25\,\text{mm}^2$. Available spaces were filled with supply decoupling caps, which are not believed to be decisive on achieving the reported results. A chip micrograph is provided in Fig. 3.42.

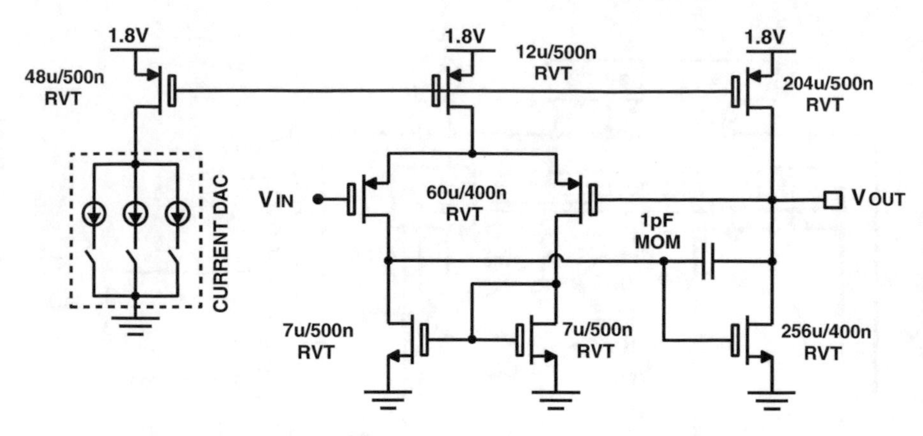

Fig. 3.41 Unity-gain voltage amplifier schematic

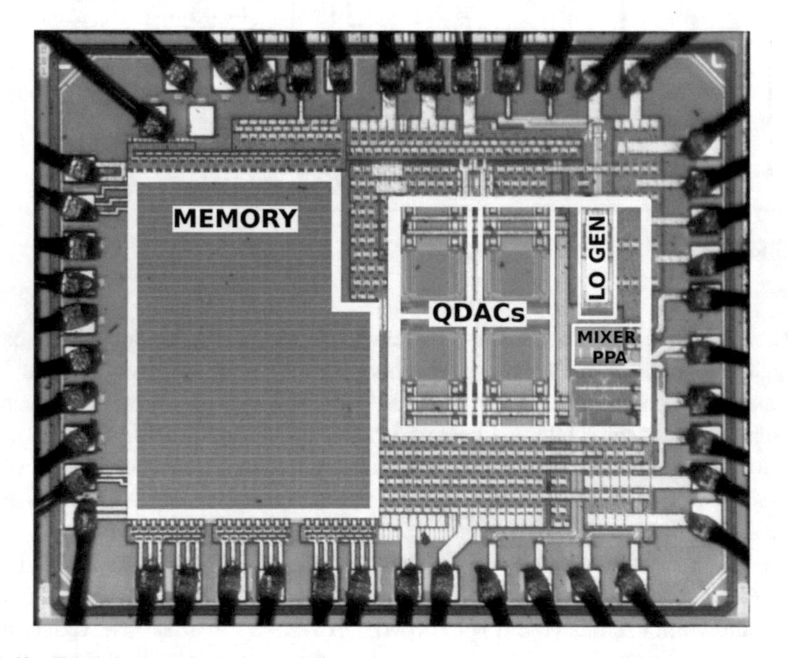

Fig. 3.42 Chip micrograph

3.4 Measurement Results

3.4.1 Measurement Setup

For this design, a custom-made PCB (FR4—four layers) is fabricated with no special features for making shorter bonding connections. The chip is directly bonded to the board using conventional wire-bonds.

To improve observability and avoid that supply coupled noise deteriorate the noise performance of the CQDAC, distinct supplies were assigned to each one of the sensitive blocks, including the CQDAC's analog supply, the PPA and the MIXER bias voltages. In total seven different supplies are used, all provided by a Agilent N6705 regulated supply source with multiple outputs.

The output stage was biased using a bias tee and a 1.8 V dedicated supply. The RF signals were directly measured using a R&S FSW-26 spectrum analyzer, while the baseband voltages were buffered with a differential active probe (TEKTRONIX 1163).

During the initial measurement phase, few samples had their output stage (PPA) damaged after repetitive test cycles, indicated by a sudden degradation of the system linearity and a large increase in the gate current seen at the PPA cascode transistor. The reason is believed to be an oxide breakdown caused by consecutive discharges of the large bias tee's inductor every time the transmitter was re-programmed and the PPA input transistor was cut (note that the PPA is DC coupled to the mixer output). This problem was solved by shunting the RF output with an extra 50 Ω resistor that worked as a discharge path for the bias tee inductance. The implied signal attenuation could be properly de-embedded from measurements and no additional samples were damaged after this measure.

Another important issue addressed during measurements was the increased phase imbalance given by the hybrid coupler used to convert the LO signal from single-ended to differential. The minimum duty-cycle configuration achieved with the LO generation block (approximately 23 %) was not sufficient to avoid the resulting overlap between the different LO phases, caused by a measured 10° phase error at 2.4 GHz. To solve the problem, the LO duty-cycle was reduced at the input (thus at $2 \cdot f_{LO}$) using a HP 8133A pulse generator, that unfortunately had a maximum lock frequency limitation of 3 GHz, implying a maximum LO frequency of 1.5 GHz. For this reason, RF measurements of the CQDAC prototype are limited to 1 GHz LO frequency. This problem was solved in the following chip by increasing the duty-cycle control range and including back-to-back inverters along the LO generation chain, as discussed in Chap. 4.

An overview of the measurement setup is shown in Fig. 3.43.

3.4.2 CQDAC Measurement Results

As pointed in Sect. 2.2.1.4, a flawless QDAC operation relies on an accurate quantification of the different accumulators involved in the charge-based operation, namely the unit capacitor C_{UNIT}, the baseband capacitance C_{BB} and the RF load capacitance C_{GS-PPA}. Errors larger than 5 % may incur considerable distortion, so checking the values of C_{UNIT} and C_{BB} post-fabrication is a key step to achieving an improved harmonic performance.

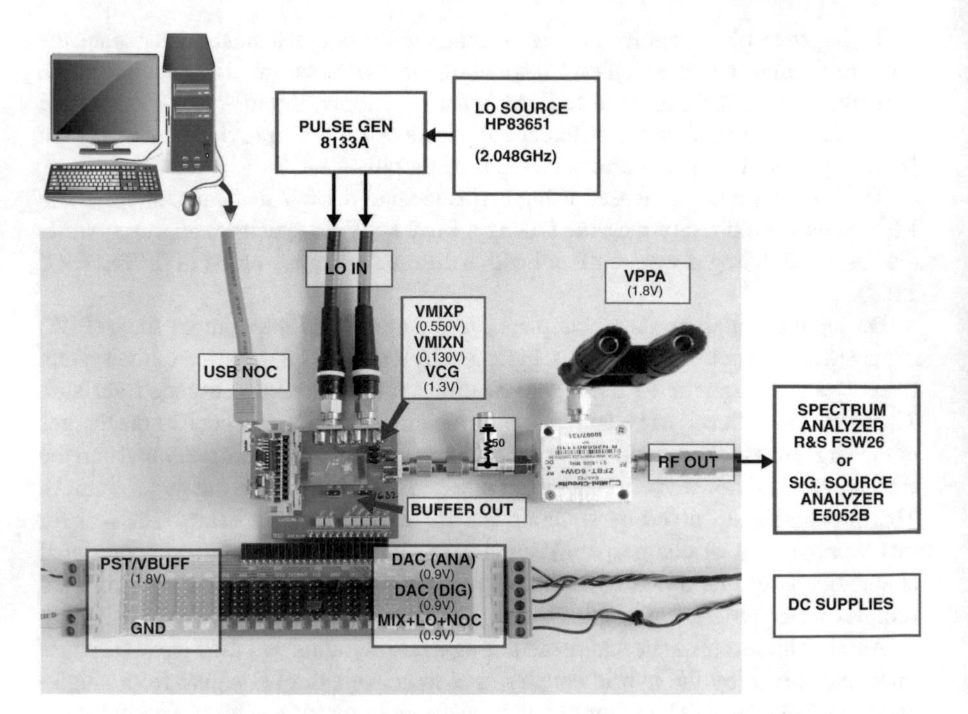

Fig. 3.43 Overview of the measurement setup

In the charge-based DAC the baseband voltage is the result of continuous charge accumulation over time. The digital input word, in turn, represents the amount of capacitance utilized to convey the required amount of charge at time "k". As a result, no direct correlation can be traced between a particular DAC input word and the instantaneous output voltage.

The only way to measure the QDAC characteristics is dynamically. By first switching the mixer OFF, two different measurements are used to define the CQDAC capacitances. To define C_{BB}, a square waveform is produced at the baseband node by repeating the same C_{DAC} value, alternately charging and discharging the baseband capacitor, as demonstrated in Fig. 3.44.

The baseband swing (V_{pp}) measured through the unity-gain buffers is combined with the DC current consumption (I_{DC}) to provide an accurate estimation of C_{BB}, as calculated by Eq. (3.24). Note that in this measurement the LO frequency (f_{LO}) has to be reduced, so that the measured signal is not attenuated by the limited bandwidth of the unity-gain buffer. With this approach an average C_{BB} of 46.2 pF could be measured.

$$C_{BB} = \frac{I_{DC}}{\frac{f_{LO}}{2} \cdot V_{pp}} \tag{3.24}$$

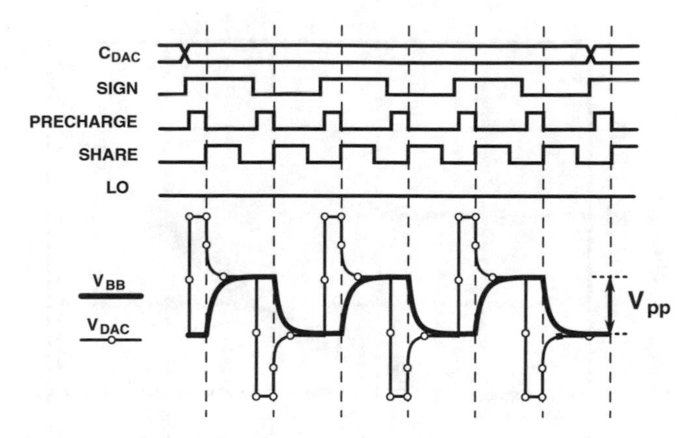

Fig. 3.44 Dynamic measurement scheme used to determine C_{UNIT} and C_{BB}

The C_{UNIT} value, in turn, is checked by sweeping the CQDAC input code so that a fixed amount of charge proportional to the number of unit cells (N) is continuously added and subtracted from C_{BB}. The unit capacitance is again calculated according to the measured voltage swing at the baseband node as follows:

$$N \cdot C_{UNIT} = \frac{2 \cdot C_{BB} \cdot V_{pp}}{V_{DD} - V_{pp}} \tag{3.25}$$

By extrapolating the first 32 values a unit capacitance of 2.44 fF is obtained with the proposed method.

Though sufficient matching should be guaranteed by design, the voltage swing at the baseband capacitor is expectedly affected by switch non-idealities such as charge injection and clock feedthrough. Increased input codes incur a larger number of devices being switched together used to convey the wanted charge, also increasing the amount of charge injected in the baseband capacitor every LO cycle. Treated as a charge-injection error (Fig. 3.45), this measurement provides valuable information about the excess charge versus input code, allowing quantification and ultimately pre-compensation of this switch non-ideality.

Measurements of the intrinsic dynamic performance of the charge-based DAC were also performed. Disabling the mixer, single-tone measurements at 128 MS/s using baseband frequencies ranging from 1 MHz to 10 MHz showed a worst-case HD3 and HD5 below −60 dBc and −68 dBc respectively, for a differential swing of 0.632 Vpp. The spectrum of a 1 MHz tone is shown in Fig. 3.46. The harmonic performance at higher sampling frequencies are shown in Fig. 3.47.

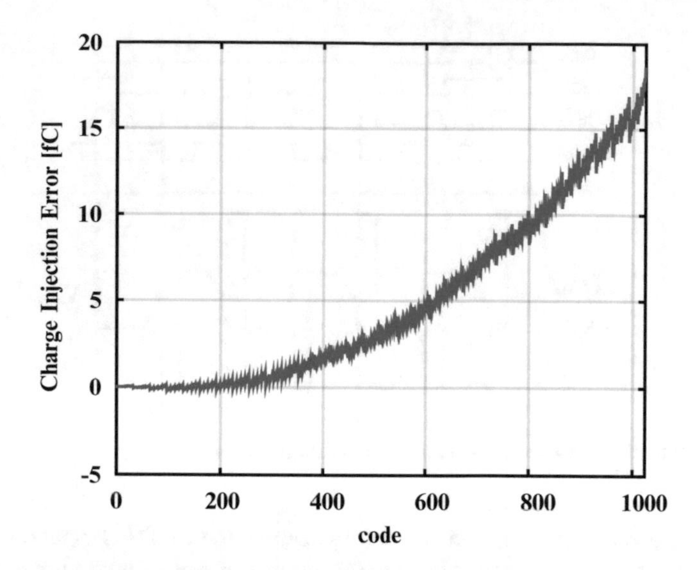

Fig. 3.45 Measured charge injection error versus input code

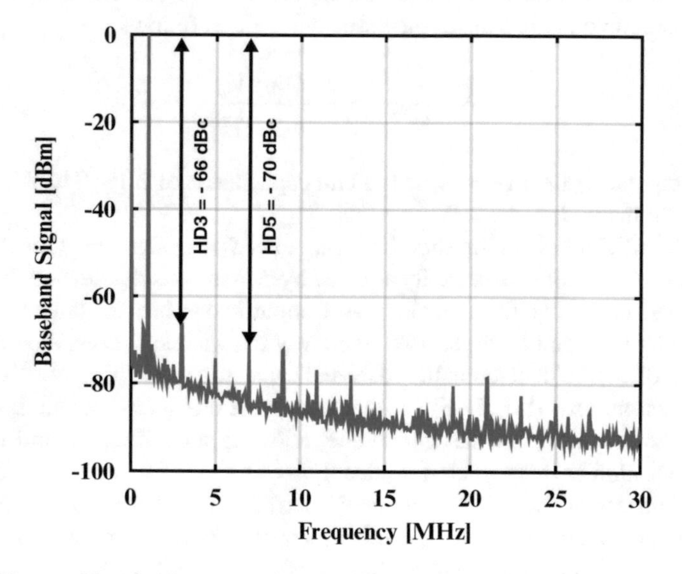

Fig. 3.46 Measured baseband spectrum for a 1 MHz single-tone, with the mixer disabled

3.4.3 CQDAC TX Measurement Results

The same test was performed with the mixer enabled, accounting the necessary RF charge to drive the PPA input capacitance. Figure 3.48 shows the measured RF spectrum for a 5 MHz single-tone sampled at 128 MS/s and transmitted at 1 GHz.

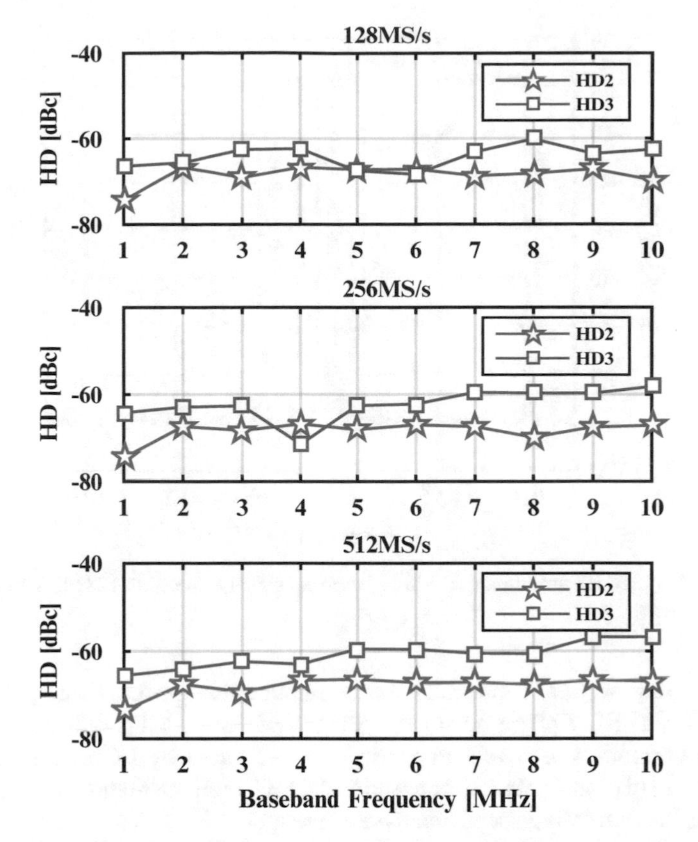

Fig. 3.47 CQDAC second and third-order harmonic distortion for different sampling frequencies

As shown in Fig. 3.49, at 7 dB backoff measurements with baseband tones ranging from 1 MHz to 20 MHz show a worst case CIM3 of −50 dBc at any baseband sampling rate. LO Leakage and I/Q image suppression are respectively better than −45 dBc and −40 dBc. Measured OP1dB of the output stage (PPA) is 8.1 dBm.

The measured relative power of the sampling aliases for multiple baseband frequencies are shown in Fig. 3.50. As expected, instead of being shaped by a sinc function as in digital transmitter architectures applying conventional DAC implementations, the charge-based transmitter features intrinsic sinc^2 alias attenuation, which yields at least 17 dB of additional suppression in the 20 MHz baseband frequency range ($F_S = 128$ MHz).

Although with the first implementation few of the most stringent out-of-band noise requirements of cellular communication systems could not be achieved, the improved out-of-band noise performance due to the architecture's intrinsic $R_{SC}C$ filtering and reduced quantization noise can be clearly visible from measurements. For the same 1 GHz modulated carrier (5 MHz CW sampled at 128 MS/s) at

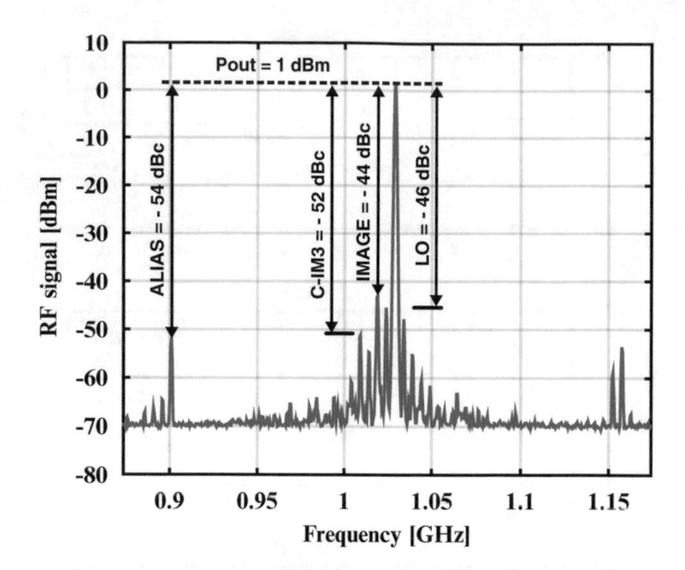

Fig. 3.48 Measured RF spectrum of a 5 MHz baseband signal sampled at 128 MS/s and transmitted at 1.024 GHz

1 dBm output power (7 dB backoff), the measured noise spectral density (including quantization and LO phase noise) is −155 dBc/Hz at 45 MHz offset (Fig. 3.51—baseband harmonics removed for clarity). Even limited by LO phase noise, this is notably 15 dB better than a conventional DAC implementation with the same number of bits, sampling speed and dynamic range.

The ACLR performance in 3G-coexistence configuration for channel bandwidths of 5, 10 and 20 MHz were also measured. With no signal pre-distortion, the measured ACLR1/ACLR2 for an RMS output power of 1 dBm and 7 dB Peak-to-Average Power Ratio (PAPR) are respectively −42/ − 47 dBc (Fig. 3.52). Due to a limited on-chip memory size, the EVM performance could not be assessed for the current implementation.

As expected, the system charge intake and its corresponding power consumption scales with signal's amplitude and frequency. For an average output power of 1 dBm, the total power consumption is mainly determined by the memory operation (28.8 mW from 0.9 V), the PPA (23 mW from 1.8 V) and the LO generation (5.7 mW from 0.9 V). The baseband circuitry consumes 12.6 mW (0.9 V), of which only 3.7 % (0.47 mW) corresponds to the system charge intake. The remaining 12.1 mW is related to the digital circuitry operation, which was not optimized in this implementation.

To conclude, a performance summary and comparison table is provided (Tables 3.1 and 3.2). As noted, the analog implementations are able to provide improved noise performance with reduced number of bits. However, when

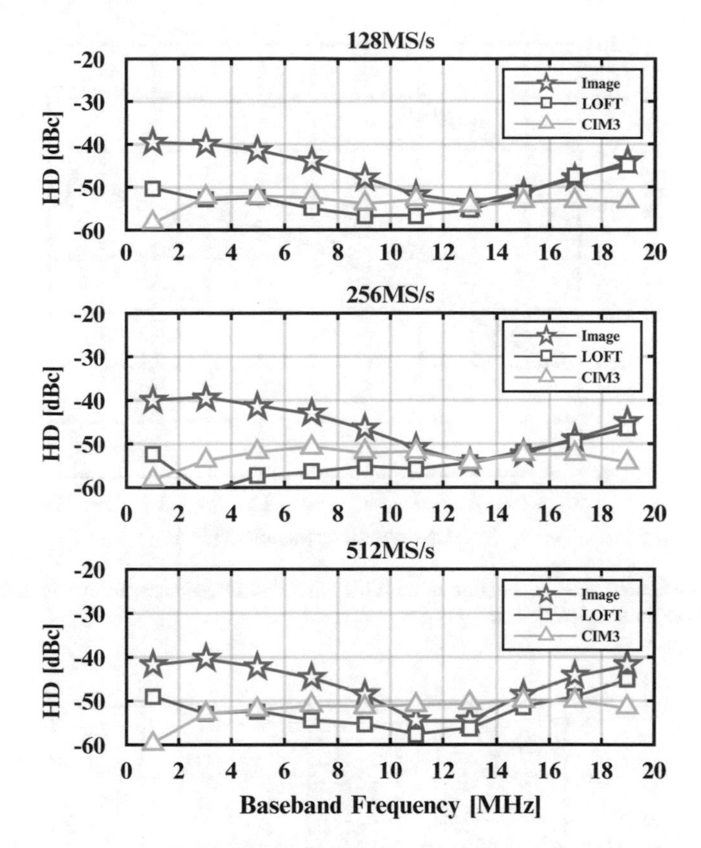

Fig. 3.49 TX spurious emission for different sampling frequencies

compared to [He10, Ros13], this implementation shows a clear improvement in area consumption—with no compromise in noise performance, mostly due to the absence of bulky RC filters in the signal path. The given digital-intensive transmitters, on the other hand, are significantly more portable and area efficient. However, with few exceptions including [Meh10]—which implements sigma-delta modulation to shift the quantization noise away from the receive band, they typically fall short in terms of out-of-band noise emission. Compared to [Elo07, Lu13, Ala14], the presented work provides a significant improvement in noise performance with a reduced DAC number of bits and sampling frequency, all thanks to its intrinsic noise filtering capability. Among both analog and digital implementations, power consumption is also among the best.

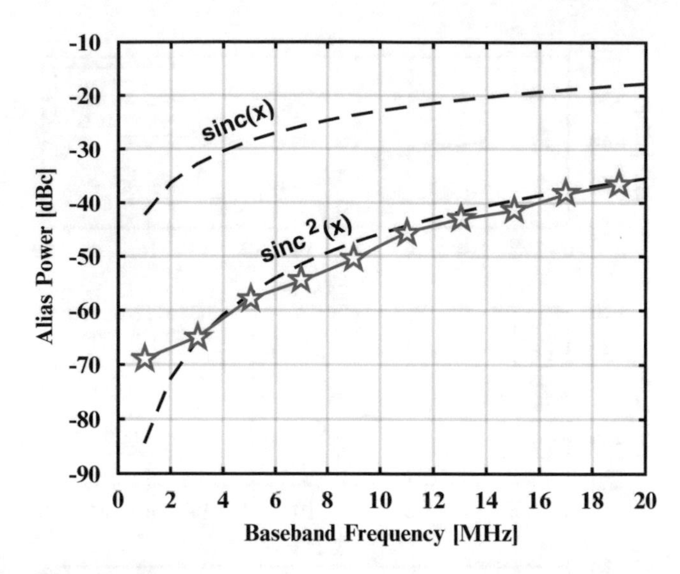

Fig. 3.50 Measured alias attenuation at multiple baseband frequencies, from 1 to 20 MHz for a 128 MS/s baseband sampling rate

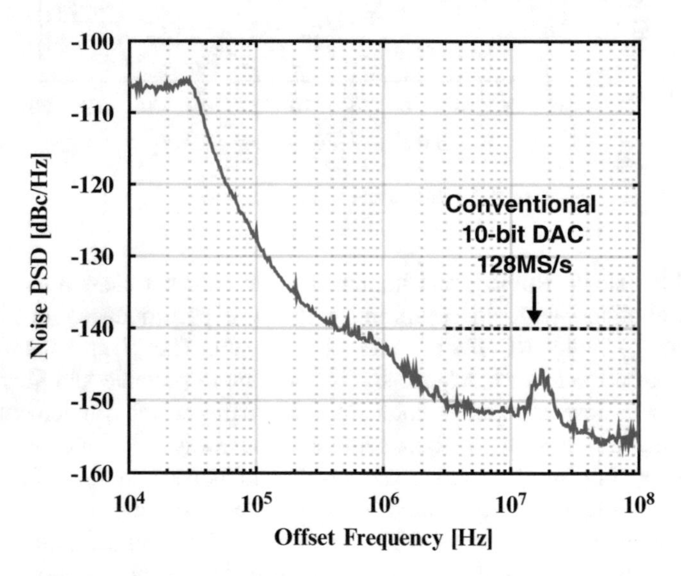

Fig. 3.51 Measured output noise for a 5 MHz, 1 dBm output power (7 dB backoff) single-tone transmitted at 1.024 GHz (baseband harmonics were removed for clarity). Both LO and quantization noise are included

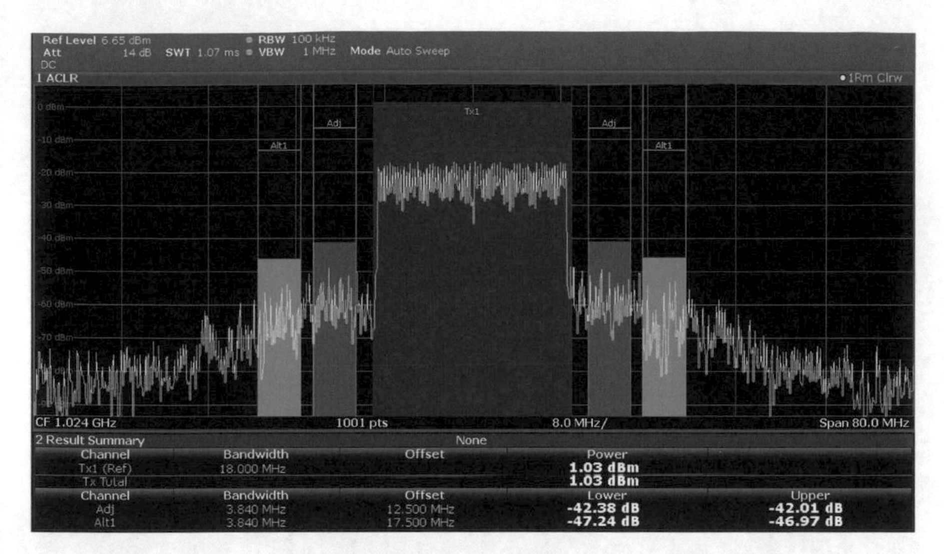

Fig. 3.52 Measured ACLR1/2 performance for 20 MHz BW

Table 3.1 Performance summary

Performance summary				
RF bandwidth	[MHz]	5	10	20
Output power	[dBm]	1	1	1
ACLR/ACLR2	[dB]	$-45/-69.5$	$-44/-50$	$-42/-47$
CIM3	[dBc]		<-50	
Noise @ 5dBm LO power	[dBc/Hz]		-155	
Offset	[MHz]		45	
Current consumption				
DAC (charge intake)		0.401	0.441	0.520
DAC (digital)		11.58	12.03	12.51
Mixer	[mA]	0.926	0.926	0.926
LO generation		6.3	6.3	6.3
PPA @ 1.8V		12.8	12.9	13.0
Memory + NOC		32	32	32
Supply voltage	[V]		0.9/1.8	
Power consumption (DAC+MIXER+PPA)	[mW]	34.65	35.27	35.96
Efficiency	[%]	3.68	3.61	3.52
Area	[mm^2]		0.25	
Process	[nm]		28	

Table 3.2 Comparison table

REF		Codega et al. JSSC 2014		Eloranta et al. JSSC 2007	Alavi et al. TMTT 2014	This work		
Architecture		ANALOG		DDRM	RFDAC	QDAC		
RF bandwidth	[MHz]	10	20	5	20	5	10	20
Max output power	[dBm]	6	6	NA	22.8	8	8	6[a]
(RF)DAC	[bits]	10		10	13		10	
Max BB Clk	[MHz]	NA		307	300		128	
Noise @Offset Power	[dBc/Hz]	<-159[b] @45 M 2.8dBm	<-158[b] @80 M 0dBm	−146 @190M −2dBm	−160 [c] @200M NA		−155 @45M 1dBm	
ACLR/ACLR2	[dBc]	−43.4/ − 55 @2.8dBm	−42.5/ − 55 @4dBm	−58/ − 61 @−2dBm	NA	−45/ − 70 @1dBm	−44/ − 50 @1dBm	−42/ − 47 @1dBm
CIM3	[dBc]	NA	−57.1 @2.3dBm	NA	NA		<-50 @1dBm	
Consumption		@2.8dBm	@4dBm	@ NA	@22.8dBm	@1dBm, LO = 1 GHz		
Modulator						0.3/11.3[d]	0.4/11.7	0.5/12.1
LO	[mW]			92 @ ANA	33	5.7	5.7	5.7
PPA				65 @ DIG	–	23	23	23
Total		97	98	157	NA	40.3	40.8	41.3
Supply	[V]	1.8		1.2	1.2		0.9/1.8	
Active area	[mm^2]	1.3		–	0.45		0.25	
Process	[nm]	55 LP		130	40		28	

[a] Maximum output power limited by maximum DAC capacitance
[b] Using Baseband reconstruction filter
[c] Static measurement (does not include quantization noise)
[d] Modulator power consumption split into (charge intake)/(digital)

3.5 Conclusion

Starting from a top-level description, in this Chapter a top-down approach was used to explore all the different aspects concerning the design and implementation of what is considered to be the first made charge-based transmitter.

In Sect. 3.2.1, the operating principles of both capacitive charge-based DAC (CQDAC) and transmitter are discussed. It is shown that the whole transmitter operation is based on two charge components, one responsible for driving the baseband capacitor, and another for the RF load. A more detailed analysis of the architecture shows that all the noise improvements discussed in Chap. 2 for a "black-box" implementation are also provided by the capacitive realization of the proposed QDAC. PNOISE simulations are also used to validate the remarkable noise filtering capabilities and the estimated noise cutoff frequency. It is shown that the minimum voltage step that can be produced by the CQDAC is proportional to the ratio between two capacitors (namely the baseband and the unit capacitor), allowing quantization noise reduction by simply adjusting C_{UNIT} and C_{BB}. According to system simulations, a better than 13-bit ENOB can be expected from the implemented C_{BB} and C_{UNIT} values of 45 pF and 2 fF, respectively.

In terms of harmonic performance, system-level simulations indicate that complete voltage settling is key to achieve a reduced harmonic distortion. If every switch along the signal path is sized to provide enough conductance so that all time-constants remain below $1/7$ of the switch ON period (approximately 7.5 ps for a 2.4 GHz LO frequency), harmonic distortion should not be limited by incomplete settling. Also regarding the switch sizes, it is demonstrated that increased charge injection implied by oversized switches can also degrade the harmonic performance. For that matter, it is shown in Sect. 3.3.1.2 that the best approach in this case is to optimize the switch sizes, and use complementary switches (NPMOS) for its reduced charge injection, LO feedthrough and improved switch conductance. The application of complementary switches is also one of the solutions found to relax the harmonic distortion implied by the switch conductance voltage-dependence observed at the passive mixer. For the same output dynamic range, the combination of a NMOS and a PMOS switch provides less variation of the equivalent switch conductance, which is further minimized by shifting the DC level of the LO signals driving the mixer switches. Further implementation details are disclosed in Sect. 3.3, with specials design remarks given for the CQDAC unit cell, PPA and mixer.

Finally, the measurement results are shown in Sect. 3.4. By switching OFF the mixer and only accounting for the required charge to drive the baseband capacitance without the influence of the RF load, the capacitive charge-based DAC showed remarkable linearity with a third-order harmonic distortion lower than -66 dBc for a differential baseband swing of 0.632 Vpp. With the mixer ON, CIM3 is always better than -50 dBc for all baseband frequencies measured, from 1 MHz to 20 MHz. It was also possible to observe the expected sinc^2 alias attenuation, leading to at least 17 dB of additional suppression at 128 MS/s. Last but not least, thanks to the intrinsic noise filtering capabilities inherently provided by the incremental charge-based operation, measured out-of-band noise at 45 MHz offset is −155 dBc/Hz, being notably 15 dB lower than the minimum noise spectral density that could possibly be attained by a conventional DAC implementation with the same number of bits and sampling speed.

Chapter 4
Resistive Charge-Based Transmitter

4.1 Introduction

In the previous chapter a discrete-time capacitive charge-based transmitter architecture was described. Using switches and capacitors, sizable packets of charge were controllably transferred from the supply to the output stage, which has the important role of increasing the signal output power before driving the subsequent Power Amplifier (PA). To provide the required output swing without degrading the signal's harmonic performance, in most cases the PPA is supplied with a higher voltage, which increases both power consumption and cost, requiring technology features such as thick-oxide devices and additional voltage regulators. Also to avoid linearity degradation, in many cases the PPA is operated in Class-A mode, which also leads to a significant impact in power consumption. In fact, more than 50 % of the total amount of power consumed in the first charge-based TX was due to the PPA contribution.

Removing as a result the PPA from the signal path would more than anything provide considerable improvements in power efficiency, but not only: First, without an additional analog block that also compresses the output signal, the charge-based transmitter linearity can be enhanced in a much more power efficient way. Second, the effectiveness of pre-distortion is also leveraged, since the PA input is driven in this case with controlled packets of charge facing no bandwidth (or linearity) limitation implied by an intermediate block. Last but not least, the PPA removal would also be a valuable opportunity to study the ability of charge-based architectures to deliver an increased amount of power to a low impedance (50 Ω) RF load, while still preserving all the benefits discussed in the previous Chapters.

As clarified in the following sections, a resistive QDAC is chosen for this implementation [Par16], since it provides the best power and area efficiencies for the target output power. Section 4.2 provides an overview of the implemented

P.E. Paro Filho et al., *Charge-based CMOS Digital RF Transmitters*, Analog Circuits and Signal Processing, DOI 10.1007/978-3-319-45787-1_4

architecture, followed by implementation details and circuit realization given in Sect. 4.3. Measurement results and Conclusions are respectively given in Sects. 4.4 and 4.5.

4.2 Architecture

An alternative charge-based transmitter architecture is presented in Fig. 4.1. For the reasons discussed above, the PPA is removed from the signal path, and a direct-launch transmitter is implemented. Four QDACs are used to provide each one of the differential I/Q components, which are converted to a single-ended output using an external balun.

In a direct-launch transmitter, the maximum RF output swing is limited to the baseband supply voltage. Even though it may look as a deal-breaker for adoption at highly-scaled technology nodes, yet decent output power levels can be achieved with current CMOS technologies. For instance, approximately 9.1 dBm output power can be provided (in differential mode) with a 0.9 V supply, which is a typical value seen among sub-40 nm technologies.

Moreover, driving the RF load without an intermediate PPA stage also means that the entire RF power has to be driven through the baseband circuitry and mixer switches. Not only the linearity constraints of each one of these blocks are tightened, but in the specific case of charge-based architectures the charge capacity of the required QDAC has to be increased to provide the wanted output power.

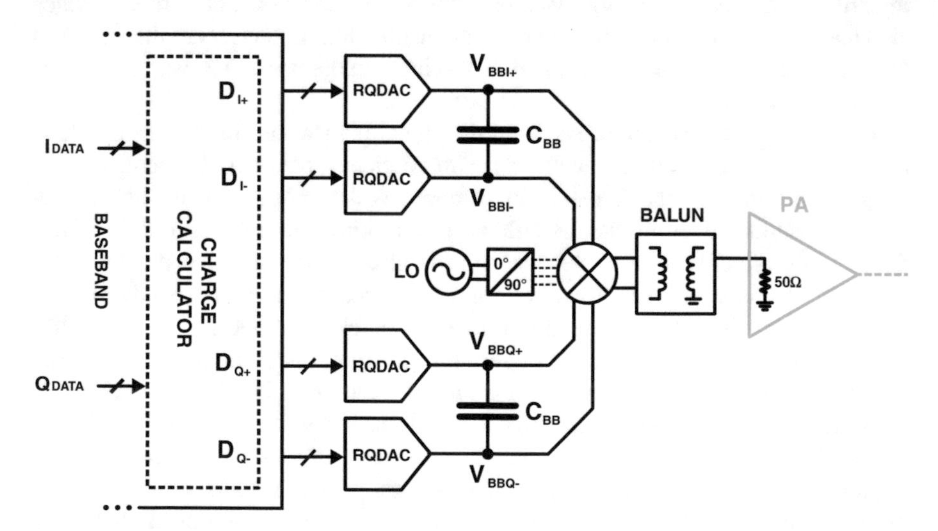

Fig. 4.1 Architecture overview of the Cartesian resistive charge-based DAC transmitter. Four QDACs provide each one of the differential I/Q components, however the RF load is now directly driven by the baseband circuitry through passive switch-based mixer, without a PPA

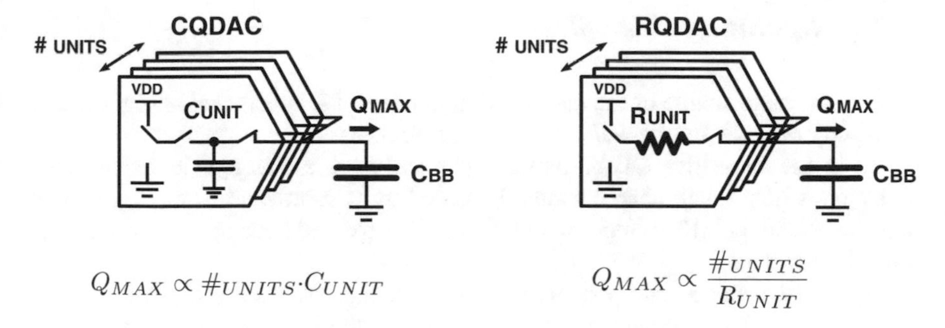

$$Q_{MAX} \propto \#_{UNITS} \cdot C_{UNIT} \qquad Q_{MAX} \propto \frac{\#_{UNITS}}{R_{UNIT}}$$

Fig. 4.2 Maximum charge capacity of both capacitive and resistive QDACs

Looking at the first implementation proposed in Chap. 3, the only way to increase the CQDAC charge capacity is by either increasing the unit capacitance (C_{UNIT}), or the total number of unit cells. In both cases, a significant impact in area consumption would be observed.

However, if instead of using unit capacitors the required QDAC is built through the parallel combination of switchable resistors, the maximum charge capacity of the resulting conductance DAC is inversely proportional to the unit resistance, as shown in Fig. 4.2. Therefore, as long as matching constraints permit, the charge capacity of the resistive QDAC can be increased by decreasing the unit resistance R_{UNIT}, leading in this case to a smaller area consumption. As further explained in Sect. 4.3.1, by means of a resistive implementation, the charge capacity required to provide a maximum output power of 7 dBm can be achieved with only $1/4$ of the area that would be otherwise necessary if a capacitive QDAC were to be used.

A conductance DAC can also leverage the system efficiency. Remember that in the previous charge-based implementation the charge convey from supply to the baseband capacitance was operated at LO speed, charging and discharging the DAC capacitance every LO cycle before the mixer switch was closed. The expressive amount of dynamic power consumed to buffer the LO signals and drive the DAC switches at high frequencies was a significant contributor to power performance degradation in the previous TX. With a resistor-based QDAC, on the other hand, the baseband voltage is changed between consecutive values by transferring charge in continuous-time, charging and discharging C_{BB} through the DAC conductance. The modulator in this case can be operated at the much lower baseband sampling frequency, relaxing switching speed and saving power.

Moreover, in the capacitive QDAC operation the DAC switching speed is forced to be an integer fraction of LO frequency, otherwise the mixer phases would overlap with the CQDAC charge convey to C_{BB}. In a conductance QDAC, on the other hand, since the charging of C_{BB} is done continuously over time, the DAC input sampling frequency can be chosen independently of the LO frequency, without affecting the charge-based TX operation. This aspect simplifies immensely when integrating the charge-based fronted to different baseband engines.

4.2.1 Operating Principles

The operating principle of the resistive charge-based TX is depicted in Fig. 4.3. For simplicity only the in-phase I/Q component is shown.

As in the capacitive QDAC-based implementation, a charge calculation block determines how much charge should be added to or subtracted from C_{BB}, so that both baseband and RF voltages may follow their expected envelopes defined by the digital input data.

Again, two charge components are involved: the baseband (Q_{BB}) and the RF (Q_{RF}) charges. The baseband charge corresponds to the incremental charge required to drive the baseband voltage V_{BB} across consecutive baseband samples ($V_{BB}[k-1]$, $V_{BB}[k]$). Using the positive in-phase (I^{+}) I/Q signal component as an example, the required Q_{BB_I+} is given by:

$$\begin{aligned} Q_{BB_I+}[k] &= 2 \cdot C_{BB} \cdot \Delta V_{BB_I+}[k] \\ &= 2 \cdot C_{BB} \cdot (V_{BB_I+}[k] - V_{BB_I+}[k-1]) \end{aligned} \tag{4.1}$$

The RF charge, in turn, stands for the amount of charge needed to drive the RF load, subtracted from the baseband capacitor every time the mixer switch is closed. It basically depends on the instantaneous output current ($I_{OUT}[k]$) and the amount of time during which the mixer switch remains closed (T_{ON}).

$$\begin{aligned} Q'_{RF_I+}[k] &= I_{OUT}[k] \cdot T_{ON} \\ &= \frac{\Delta V_{RF}[k]}{R_{LOAD}} \cdot \frac{DC_{LO}}{F_{LO}} \end{aligned} \tag{4.2}$$

Non-negligible parasitics (C_{PAR}—Fig. 4.4) including ESD/PCB loading of the output node can also be accounted on the RF charge calculations. Different from

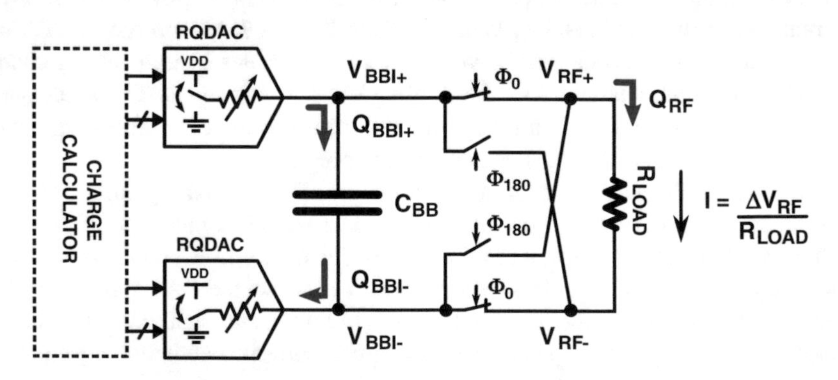

Fig. 4.3 Resistive charge-based transmitter operating principle

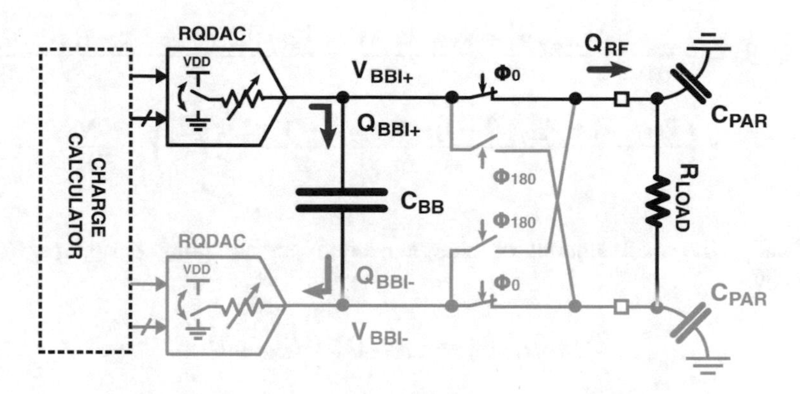

Fig. 4.4 Simplified schematic showing the parasitic output capacitance C_{PAR}

the previous TX however, the impact of C_{PAR} is expected to be reduced in this case since the RF load should be dominated by the 50 Ω PA input impedance.

$$Q'_{RF_I+}[k] = \frac{\Delta V_{RF}[k]}{R_{LOAD}} \cdot \frac{DC_{LO}}{F_{LO}} + \left(V_{BB_I+}[k] - V_{BB_Q-}[k]\right) \cdot C_{PAR} \tag{4.3}$$

The total RF charge (Q_{RF}) needed per sampling period is therefore given by Eq. (4.4). C_{PAR} is neglected for simplicity.

$$\begin{aligned} Q_{RF_I+}[k] &= Q'_{RF}[k] \cdot \frac{F_{LO}}{F_S} \\ &= 2 \cdot \frac{\Delta V_{RF_I+}[k]}{R_{LOAD}} \cdot \frac{DC_{LO}}{F_{LO}} \cdot \frac{F_{LO}}{F_S} \\ &= 2^1 \cdot \frac{(V_{BB_I+}[k] - V_{BB_I-}[k])}{R_{LOAD}} \cdot \frac{DC_{LO}}{F_S} \end{aligned} \tag{4.4}$$

As in Eq. (3.4), accuracy is improved when the instantaneous charging and discharging of C_{BB} is accounted in the Q_{RF} calculations, done in this case by considering the average value between samples, as demonstrated in Eq. (4.5):

[1]The factor 2 is introduced because the RF load is connected twice to the baseband node per LO cycle.

$$Q_{RF_I+}[k] = \frac{2}{R_{LOAD}} \cdot \left(\frac{V_{BB_I+}[k] + V_{BB_I+}[k-1]}{2} - \frac{V_{BB_I-}[k] + V_{BB_I-}[k-1]}{2} \right) \cdot \frac{DC_{LO}}{F_S}$$
$$= \left(\frac{V_{BB_I+}[k] + V_{BB_I+}[k-1] - V_{BB_I-}[k] - V_{BB_I-}[k-1]}{R_{LOAD}} \right) \cdot \frac{DC_{LO}}{F_S} \tag{4.5}$$

Finally, the total amount of charge needed per sampling period ($1/F_S$) is given by:

$$Q_{TOTAL_I+}[k] = Q_{BB_I+}[k] + Q_{RF_I+}[k] \tag{4.6}$$

The charge calculations are realized at baseband sampling rate (F_S). For every baseband sample the required $Q_{TOTAL}[k]$ is recalculated, and the corresponding DAC conductance is selected.

4.2.2 RQDAC Operation

A simplified block diagram of the resistive charge-based DAC is shown in Fig. 4.5. Through the parallel combination of switchable unit resistances, the resistive DAC provides a linearly increasing conductance proportional to the DAC input code.

For every digital input sample, the charge calculation block determines how much charge should be transferred to the baseband capacitance during the following sampling period. The charge convey between the supply and C_{BB} happens in

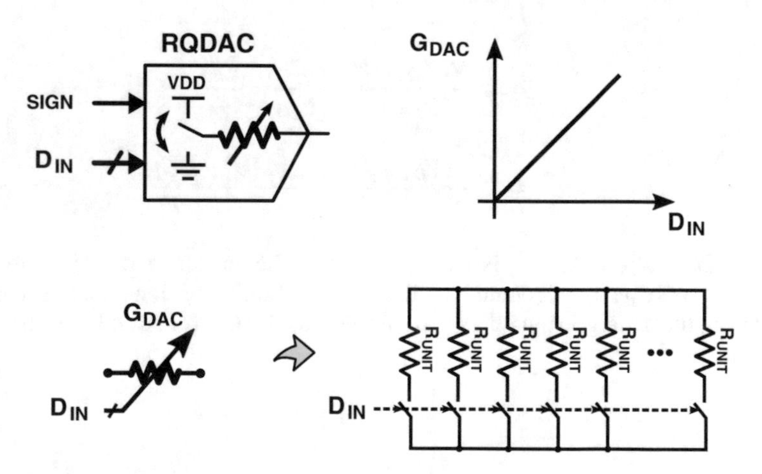

Fig. 4.5 Simplified diagram of the RQDAC implementation

continuous-time, in the form of a DAC current that is sinked into or drained from C_{BB}.

$$I_I + [k] = \frac{Q}{\Delta T_S} = Q_{TOTAL_I+}[k] \cdot F_S \tag{4.7}$$

If the net amount of charge in the system should increase during the given sampling period ($Q_{TOTAL}[k] > 0$), the DAC conductance is connected to V_{DD} as defined by the control signal SIGN shown in Eq. (4.8). Otherwise, charge is drained from C_{BB} to the ground by connecting the DAC conductance to G_{ND}. The instantaneous supply voltage to which $G_{DAC}[k]$ is connected to at time k is defined by V_{REF}, as shown in Eq. (4.9).

$$SIGN_{I+}[k] = \begin{cases} 0, & \text{if } Q_{TOTAL_I+}[k] \geq 0, \\ 1, & \text{otherwise.} \end{cases} \tag{4.8}$$

$$V_{REF_I+}[k] = \begin{cases} V_{DD}, & \text{if } SIGN_{I+}[k] = 0, \\ G_{ND}, & \text{otherwise.} \end{cases} \tag{4.9}$$

The instantaneous DAC conductance $G_{DAC}[k]$ is hence calculated considering the required output current and the voltage drop across the DAC terminals (assuming C_{BB} pre-charged to previous baseband voltage $V_{BB_I+}[k-1]$).

$$\begin{aligned} G_{DAC_I+}[k] &= \frac{I_I + [k]}{\Delta V_{RDAC_I+}[k]} \\ &= \frac{Q_{TOTAL_I+}[k] \cdot F_S}{V_{REF_I+}[k] - V_{BB_I+}[k-1]} \end{aligned} \tag{4.10}$$

Finally, the number of DAC elements required at time k is obtained by dividing the calculated $G_{DAC_I+}[k]$ by the unit conductance, given by the inverse of the RQDAC unit resistance (R_{UNIT}).

$$D_{IN_I+}[k] = \frac{G_{DAC_I+}[k]}{G_{UNIT}} = G_{DAC_I+}[k] \cdot R_{UNIT} \tag{4.11}$$

A timing diagram of the RQDAC operation is shown in Fig. 4.6. Once the correct DAC conductance is selected, the baseband voltage starts increasing following the continuous-time exponential RC charging of C_{BB} from $V_{BB_I+}[k-1]$ to $V_{BB_I+}[k]$, slightly discharging every time the mixer switch is closed. Different from the previous capacitive QDAC implementation, the DAC unit cells are now switched at baseband sampling speed, implying a considerable reduction in the amount of power consumed to operate the DAC circuitry.

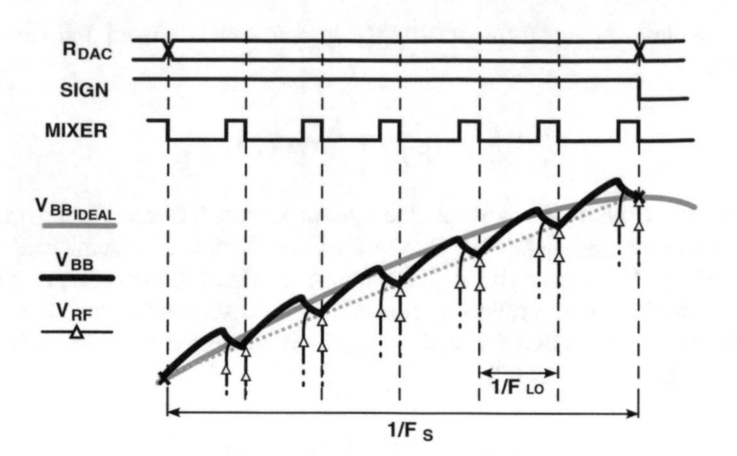

Fig. 4.6 RQDAC TX timing diagram

In it's simplest form, the DAC conductance calculation is performed assuming a fixed voltage drop ($V_{REF}[k] - V_{BB_I+}[k-1]$) across the DAC terminals during the whole sampling period. However, as one may guess this assumption is not completely valid since the charging and discharging of C_{BB} has a direct impact on the instantaneous RQDAC output current. Though satisfactory linearity can be achieved using the simplified Eq. (4.10), the harmonic performance can be further improved by considering the exponential RC charging of C_{BB}, as shown in Eq. (4.12).

$$G_{DAC_I+}[k] = -\log\left(1 - \frac{Q_{TOTAL}[k]}{C_{BB}\cdot(V_{REF_I+}[k] - V_{BB_I+}[k-1])}\right)\cdot C_{BB}\cdot F_S \tag{4.12}$$

4.2.3 *Noise and Alias Performance*

4.2.3.1 Intrinsic RC Noise Filtering

The analysis of the resistive charge-based TX clearly reveals the existence of an equivalent single-order RC filter in the signal path, as shown in Fig. 4.7.

The strong correlation between the instantaneous DAC conductance and the required amount of charge (Q_{TOTAL}) yields a signal-dependent RC filter whose bandwidth is automatically adjusted to provide the wanted voltage excursions at the transmitter output. In this way, the transmit signal is propagated through the charge-based architecture unattenuated, but non-correlated noise (e.g. thermal, supply-coupled) is intrinsically filtered by the architecture with an equivalent cutoff frequency given by the average value of the DAC conductance.

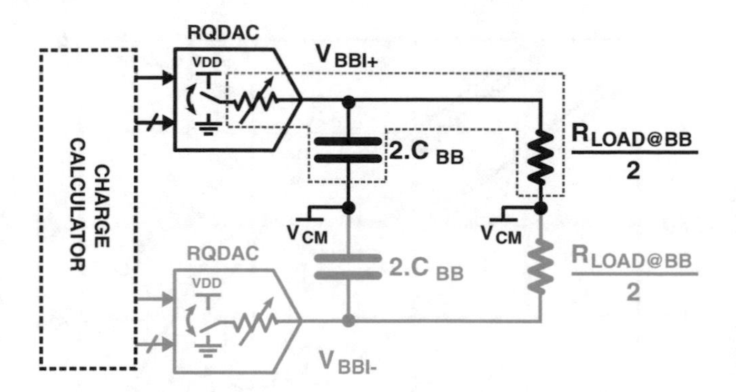

Fig. 4.7 Simplified schematic including the equivalent RF impedance seen from the baseband node

However, in the specific case of a direct-launch transmitter the RF load is connected to C_{BB} every LO cycle, also impacting the equivalent RC time-constant observed at the baseband node. Assuming ideal mixer switches, the equivalent impedance seen from baseband node toward the RF load is given by Eq. (4.13),

$$R_{LOAD@BB} = \frac{R_{LOAD}}{2 \cdot DC_{LO}} \tag{4.13}$$

leading to an equivalent noise cutoff frequency given by:

$$f_{-3\,dB} = \frac{(G_{DAC_{AVG}}) + \dfrac{4 \cdot DC_{LO}}{R_{LOAD}}}{4\pi C_{BB}} \tag{4.14}$$

As deduced from Eq. (4.14), low impedance RF loads may therefore overweight the resistive DAC average conductance, consequently shifting the equivalent f_{-3dB} to higher frequencies and reducing the noise filtering effect.

To demonstrate the impact of the RF load in the noise cutoff frequency, Fig. 4.8 shows the example PNOISE simulation for a 8 MHz 700 mV peak-peak (1.4 Vpp differential) baseband single-tone, with a baseband capacitor of 100 pF. The impact of a low impedance (50 Ω) RF load on the noise f_{-3dB} is clearly noted.

Figures 4.9 and 4.10 shows the noise f_{-3dB} dependence on the baseband signal's amplitude and frequency, respectively. As in the CQDAC implementation, the required DAC conductance—and hence the noise cutoff frequency—increases significantly when the baseband amplitude approaches the supply voltage. For the same reason a similar trend is observed in Fig. 4.10, where the increasing baseband frequencies also implies larger cutoff frequencies. If it was not for the pole shift produced by the RF load, the RQDAC implementation would also provide noise cutoff frequencies well below the actual baseband signal's.

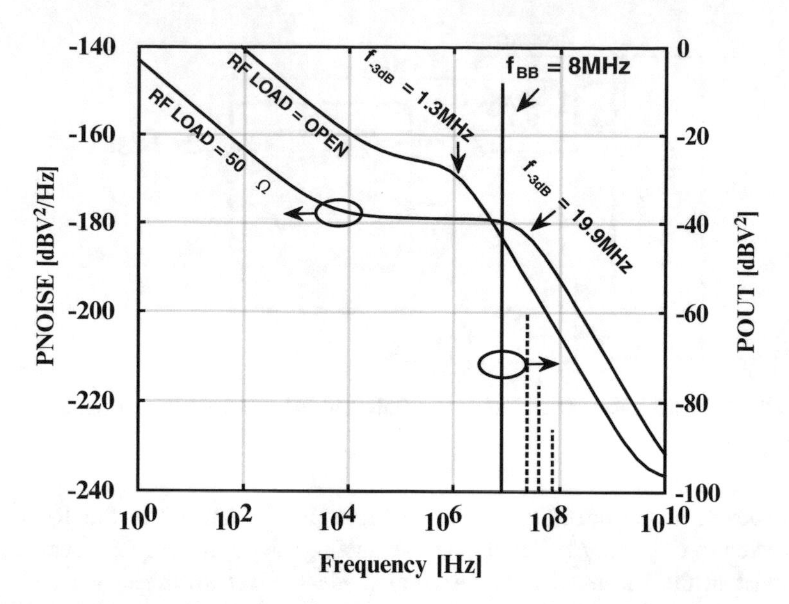

Fig. 4.8 PNOISE simulation of a 700 mVpp, 8 MHz single-tone sampled at 500 MS/s

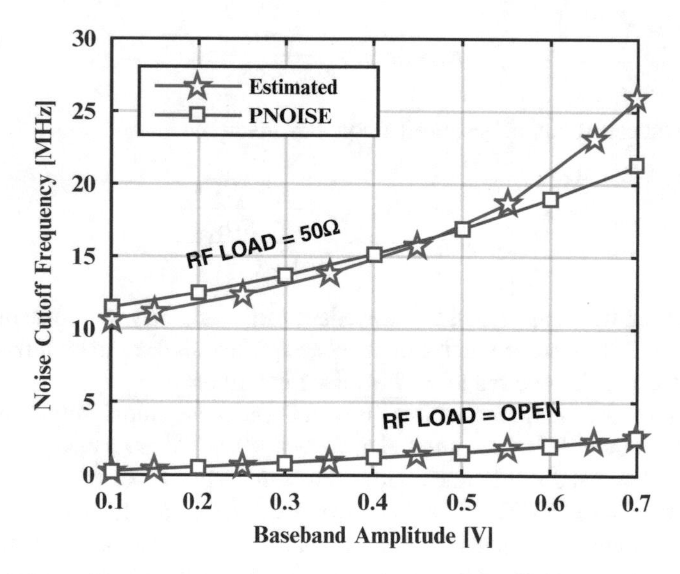

Fig. 4.9 Noise cutoff frequency versus baseband amplitude for a 8 MHz single-tone sampled at 500 MS/s, and a baseband capacitance of 150 pF

Nevertheless, the increase in the baseband node conductance can still be counterbalanced with an increase in the baseband capacitance C_{BB}, as noted in Fig. 4.11. Depending on noise performance or output power requirements, the baseband capacitor can be reduced in order to optimize the resulting area consumption.

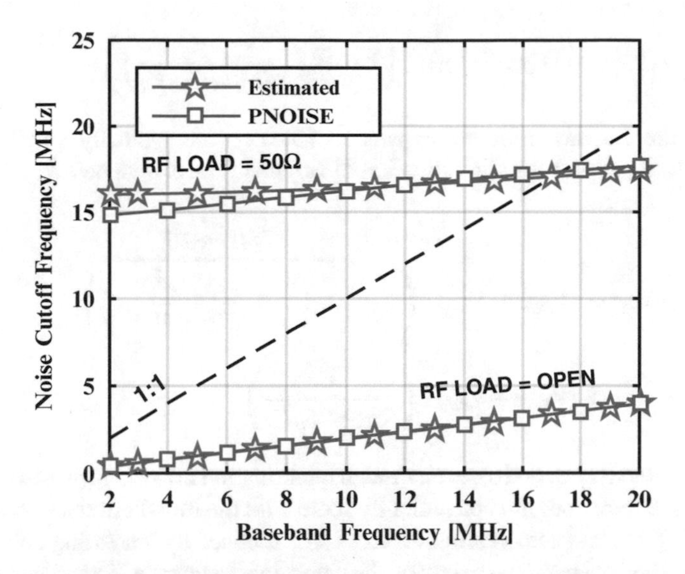

Fig. 4.10 Noise cutoff frequency versus baseband frequency for a 500 mVpp single-tone sampled at 500 MS/s, and a baseband capacitance of 150 pF

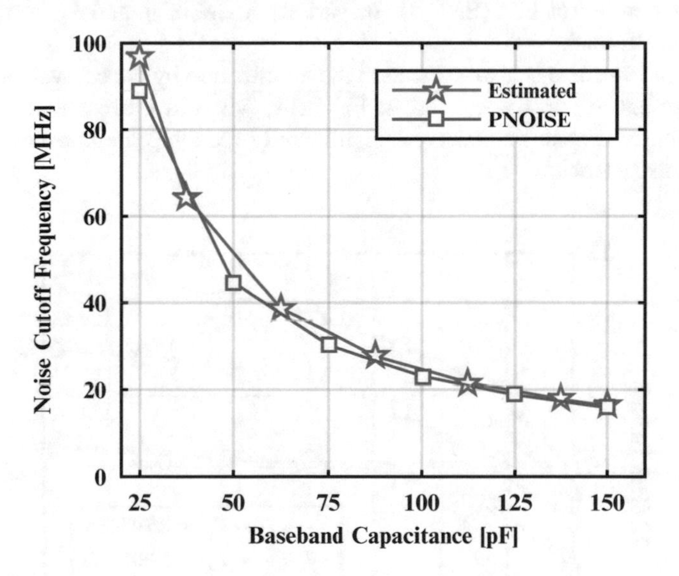

Fig. 4.11 Noise cutoff frequency versus baseband capacitance for a 500 mVpp, 8 MHz single-tone sampled at 500 MS/s

4.2.3.2 Quantization Noise

In the resistive DAC implementation, the minimum voltage step that can be resolved at the baseband capacitor C_{BB} is proportional to the minimum DAC conductance, in this case determined by the unit cell resistance (R_{UNIT}) as shown in Eq. (4.15).

$$\Delta V_{MIN} = (V_{REF} - V_{BB}) \cdot \left(1 - \exp\left(\frac{-1}{R_{UNIT}C_{BB}F_S}\right)\right) \tag{4.15}$$

Using the Maclaurin series expansion [Ste12], the typically small exponent $1/R_{UNIT}C_{BB}F_S$ allows the simplification of Eq. (4.15) as shown below:

$$\begin{aligned} \Delta V_{MIN} &= (V_{REF} - V_{BB}) \cdot \left(1 - \overbrace{\left(1 + x + \frac{x^2}{2!} + \frac{x^3}{3!} \dots\right)}^{x=\frac{-1}{R_{UNIT}C_{BB}F_S}}\right) \\ &\approx (V_{REF} - V_{BB}) \cdot \left(\frac{1}{R_{UNIT}C_{BB}F_S}\right) \end{aligned} \tag{4.16}$$

As indicated by Eq. (4.16), even though quantization noise in the resistive charge-based architecture does not conveniently scale with the ratio between two capacitors as in the CQDAC implementation, it can still be reduced by increasing either the unit resistance, the baseband capacitor or the sampling frequency. This assumption can be validated with simulations as seen in Figs. 4.12 and 4.13, showing the RQDAC equivalent number of bits (ENOB) for various R_{UNIT}, C_{BB} and F_S combinations, without the RF load.

Quantization noise is also reduced in the architecture by the equivalent RC filter. This advantage is clearly noticed in Fig. 4.14, where a clear attenuation of the quantization noise can be observed, significantly reducing it's impact on the out-of-band noise emission.

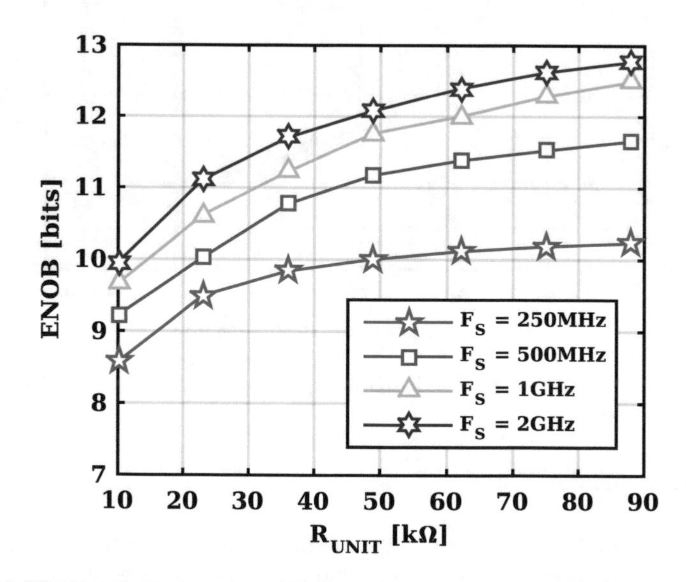

Fig. 4.12 RQDAC equivalent number of bits versus unit resistance (R_{UNIT}) at multiple sampling frequencies, for a baseband capacitance of 100 pF

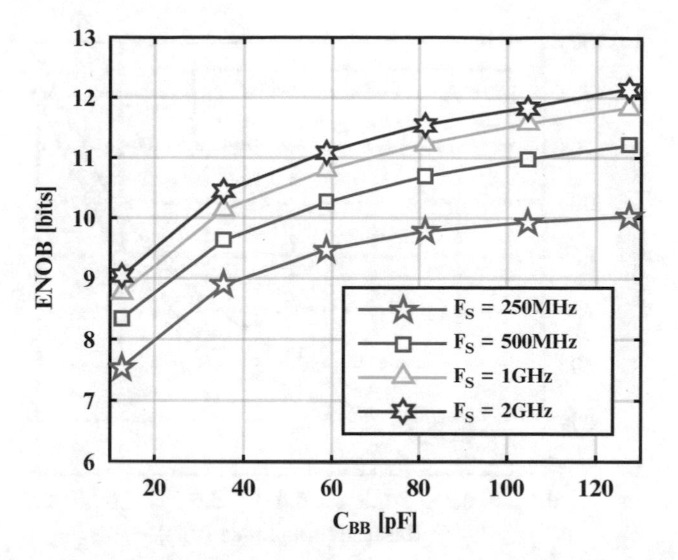

Fig. 4.13 RQDAC equivalent number of bits versus baseband capacitance (C_{BB}) at multiple sampling frequencies, for a unit resistance of 25 kΩ

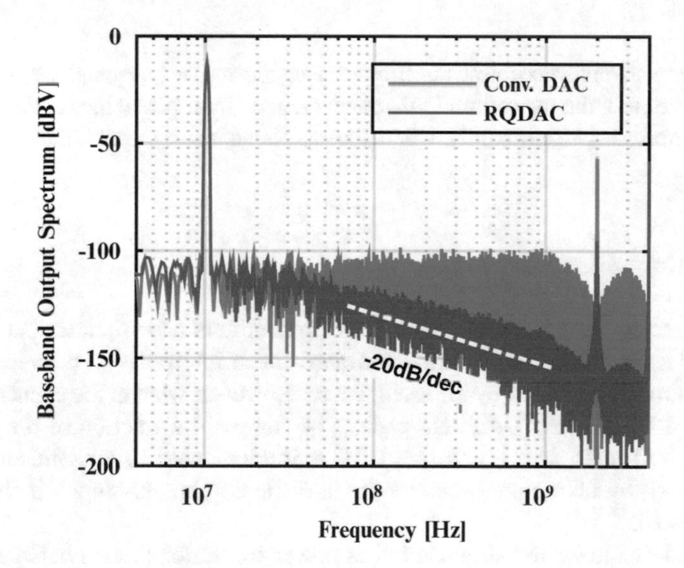

Fig. 4.14 Example baseband output spectrum showing the RQDAC transmitter quantization noise filtering

Similar to the capacitive QDAC TX, the RQDAC number of elements is not determined by quantization noise requirements. Instead, the QDAC size is defined according to the maximum DAC conductance required to provide the wanted baseband swing and frequency. Figure 4.15 shows the required RQDAC number

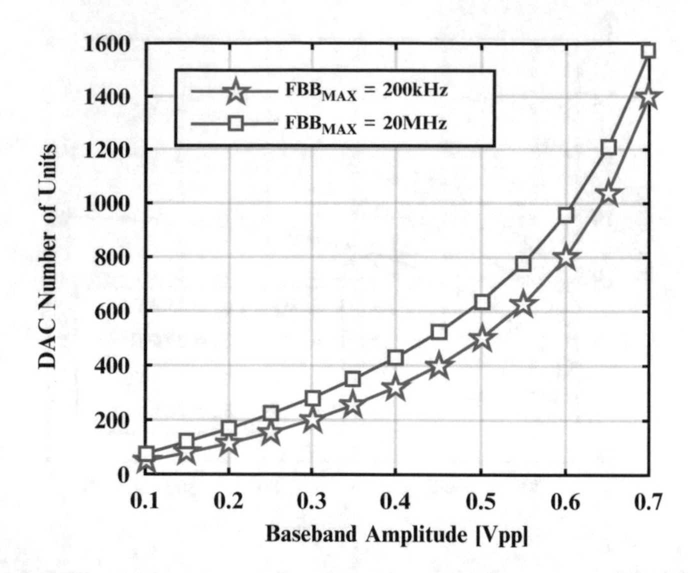

Fig. 4.15 Required number of DAC elements versus signal amplitude for two baseband maximum frequencies

of bits for various maximum amplitudes considering a baseband capacitance of 100 pF. As noted, the maximum DAC conductance—and hence the required number of bits—increases significantly when the baseband swing approaches the supply voltage.

4.2.3.3 sinc^2 Alias Attenuation

As discussed in Sect. 2.2.1.1, charge-based transmitters have their output spectrum shaped by a sinc^2 function, owing to the quasi-linear interpolation between samples that is inherently provided by the architecture. However, what in the capacitive DAC is achieved by delivering discrete packets of charge at a fraction of the sampling period (at LO rate), in the resistive DAC is implemented in a significantly more efficient way, by taking advantage of the intrinsic continuous-time RC charging of C_{BB} (Fig. 4.6).

Figure 4.16 shows the simulated alias power for various ratios between the LO and sampling frequency. As expected, when compared to a conventional RFDAC implementation, more than 30 dB of additional alias attenuation is provided in this example. Moreover, since the baseband interpolation is inherently provided the RQDAC operation, the improved alias attenuation can be achieved at virtually any sampling speed.

On top of the significant amount of power saved by running the DAC switches at a lower speed, the resistive DAC architecture also allows the detachment between the sampling and LO frequencies, increasing flexibility.

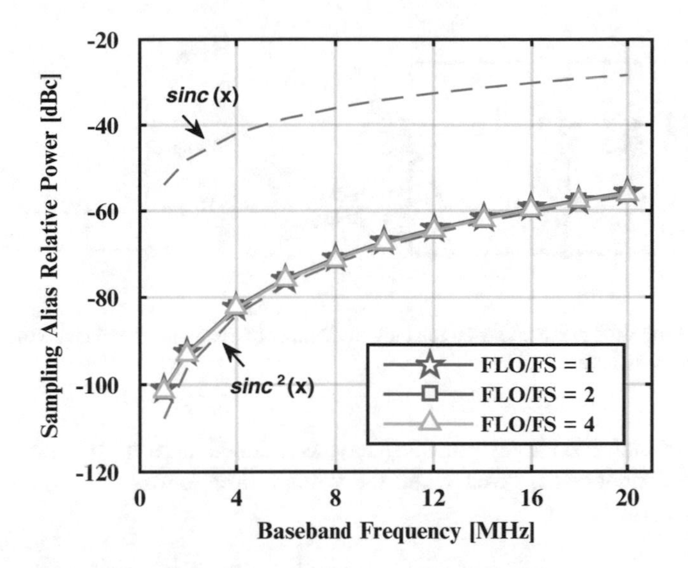

Fig. 4.16 Alias attenuation for different baseband sampling frequencies

4.2.4 *Harmonic Performance*

Without the PPA in the transmitter signal path, a relevant contributor to signal distortion is removed. However, the PPA removal also means that more power is pushed through the baseband circuitry, making it even more challenging to provide the required linearity. Since both C_{BB} and the RF load are expected to have marginal impact on the transmitter harmonic performance, the only two components left that can possibly degrade the signal integrity are respectively the RQDAC switches and the mixer switches, analysed independently.

4.2.4.1 RQDAC Switch

The simplified schematic of the resistive QDAC (Fig. 4.17) shows the unit cell consisted of a fixed unit resistance (R_{UNIT}) and a selection MOS switch. The reduced voltage dependence typical of poly-silicon integrated resistors (in contrast with diffused or well resistors) makes the MOS selection switch ON-conductance the major contributor to signal distortion deriving from the RQDAC.

For a qualitative analysis, the switch is assumed in triode operation at all times, in which case the drain-source conductance (g_{ds}) is simplified as:

$$g_{ds} = \mu_n C_{OX} \frac{W}{L} \left(V_{OV} - V_{DS} \right) \quad \text{(}V_{DS} \to \Delta V_{SW}\text{)} \tag{4.17}$$

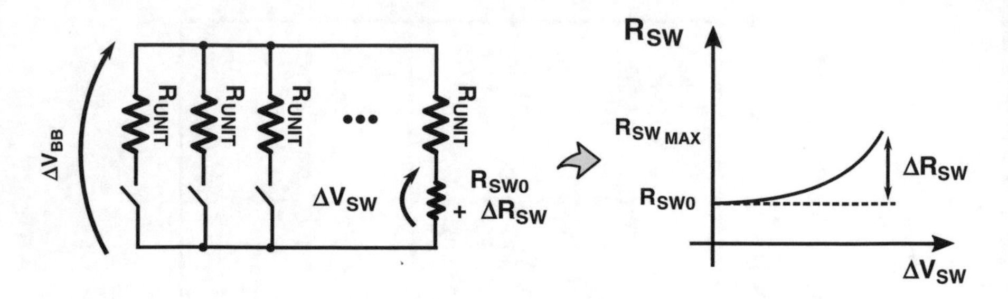

Fig. 4.17 Simplified schematic used to study the impact of the unit cell switch on the transmitter harmonic performance

The nominal switch resistance (R_{SWO}) is defined here as the inverse of the maximum conductance, given when the voltage drop across the switch terminals (ΔV_{SW}) is zero:

$$R_{SWO} = \frac{1}{\mu_n C_{OX} \frac{W}{L} (V_{OV})} \tag{4.18}$$

The maximum resistance ($R_{SW_{MAX}}$), on the other hand, is calculated when the voltage drop is maximum, defining a total switch resistance variation (ΔR_{SW}):

$$\Delta R_{SW} = \frac{\Delta V_{SW}}{\mu_n C_{OX} \frac{W}{L} V_{OV} (V_{OV} - \Delta V_{SW})} \tag{4.19}$$

The ΔR_{SW} relative to the total unit cell resistance ($R_{SW} + R_{UNIT}$) is thus given by:

$$\frac{\Delta R_{SW}}{R_{SWO} + R_{UNIT}} = \frac{\Delta V_{SW}}{(V_{OV} - \Delta V_{SW}) \left(1 + \frac{R_{UNIT}}{R_{SWO}}\right)} \tag{4.20}$$

where ΔV_{SW} can be calculated as a fraction of the RQDAC output swing (ΔV_{BB}).

The analysis of Eq. (4.20) leads to the following conclusions: First, the impact of the switch voltage dependence can be reduced by either increasing the switch overdrive (V_{OV}), or reducing the baseband voltage swing, which is not convenient in this case since the output power should be maximized. Second, the relative switch resistance variation is also decreased when R_{SWO} is made sufficiently smaller than the fixed unit resistance R_{UNIT}, translated into a larger R_{UNIT}/R_{SWO} ratio.

Using different switches sizes and a 25 kΩ unit resistance, the harmonic performance versus R_{UNIT}/R_{SWO} is studied. Figure 4.18 shows the spectrum of a single-ended RQDAC implementation, where the increase of R_{UNIT}/R_{SWO} from 6 to 20 provides the respective HD2 and HD3 improvements of 25 and 18 dB. The same

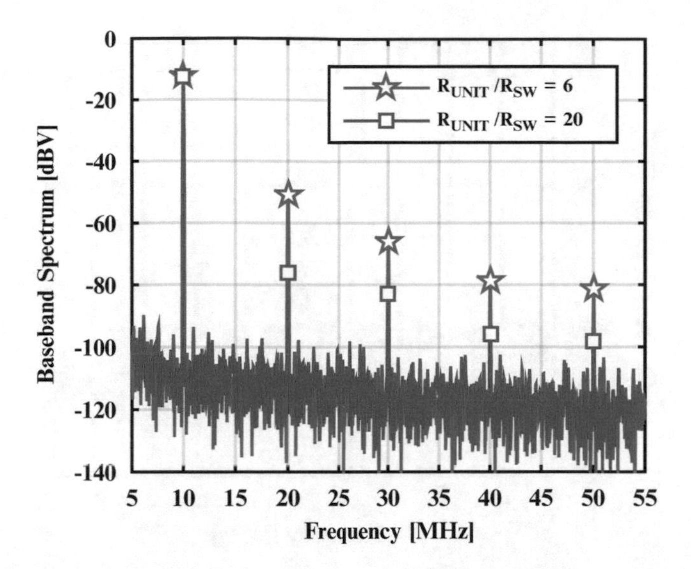

Fig. 4.18 Example baseband output spectrum for two implementations with different switch resistances

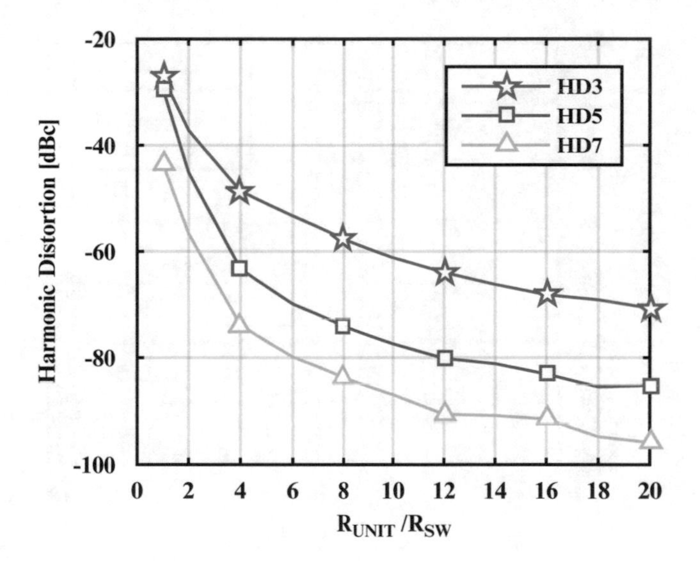

Fig. 4.19 Harmonic distortion versus RQDAC switch resistance

trend is observed in Fig. 4.19, where the dominant odd harmonics are analysed for multiple resistance ratios.

The control switch OFF-resistance is another important parameter with respect to linearity since it directly impacts on charge leakage from C_{BB}. As demonstrated in Figs. 4.20 and 4.21, the transmitter harmonic performance is improved when

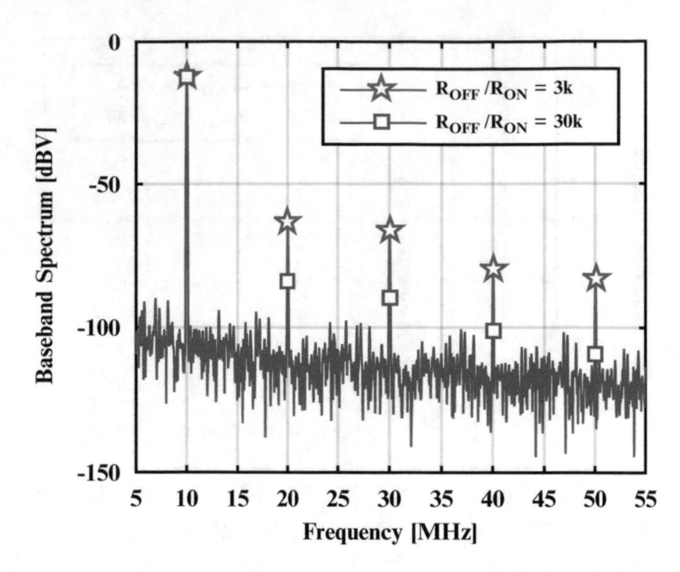

Fig. 4.20 Example baseband output spectrum for two implementations with different R_{OFF}/R_{ON} ratios

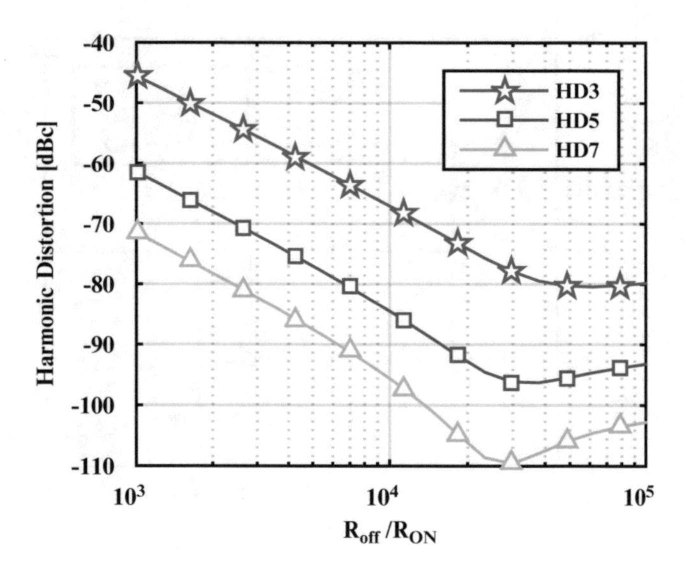

Fig. 4.21 Harmonic distortion versus R_{OFF}/R_{ON}

the ratio R_{OFF}/R_{ON} is increased. Figure 4.20 shows the impact of two different R_{OFF}/R_{ON} ratios on the baseband spectrum, where the increase from 3 to 30 k incurs 20 and 24 dB improvements on HD2 and HD3 respectively. Despite the larger conductance attained with the same overdrive, ultra-low Vth MOS switches may not be appropriate in this case due to a typically poorer R_{OFF}/R_{ON} ($\sim$10 k in a sub-45 nm

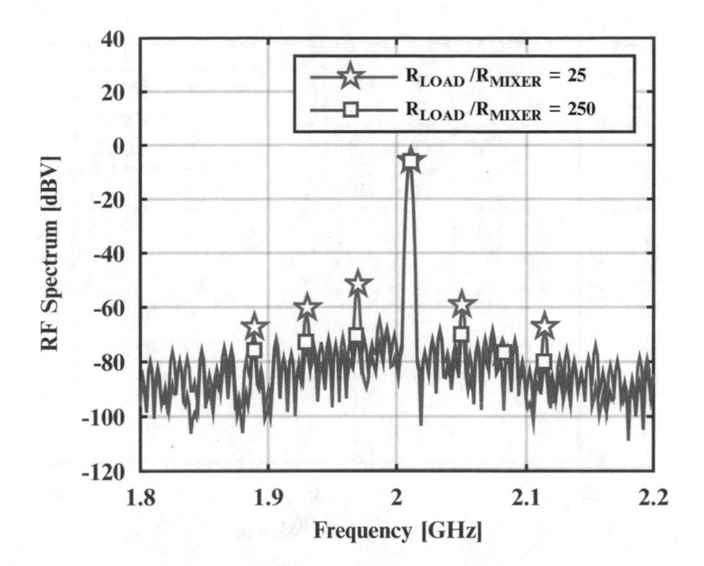

Fig. 4.22 RF output spectrum for two example implementations with different switch resistances

node, versus ~100 k provided by a regular Vth minimum-length MOS switch of the same technology).

4.2.4.2 Mixer Switches

The same reasoning is applied to the mixer switches in attempt to assess their contribution to signal distortion. Different from the resistive DAC control switches, the mixer switches do not have a fixed overdrive, implying a inevitable ON-resistance voltage dependence to the baseband voltage V_{BB}.

Figure 4.22 shows the RF spectrum using different switch sizes with equivalent R_{MIXER} of 2 and 0.2 Ω (50 Ω RF load). The harmonic performance for various switch resistances is shown in Fig. 4.23. In this analysis, complementary switches were used for their lower ΔR_{MIXER}, charge injection and clock feedthrough.

The harmonic distortion produced in this example demonstrates that especially when low impedance RF loads are used, the harmonic performance of the resistive charge-based transmitter is notably dominated by the mixer non-linearity. Again, the impact of the switch ON-resistance modulation can only be reduced by decreasing the impact of ΔR_{MIXER} on the total RF impedance ($R_{LOAD} + 2R_{MIXER}$), also achieved by reducing the mixer switch resistance.

However, though transmission loss is also decreased when R_{MIXER} is reduced, the benefits of having low resistance mixer switches are hampered by excessive area and LO-driving power consumption, and parasitic capacitance loading of the output node, ultimately degrading the harmonic performance.

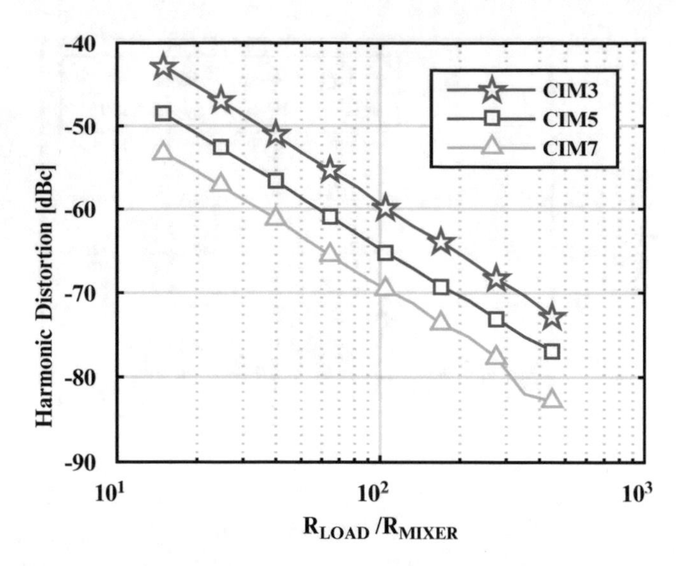

Fig. 4.23 Harmonic distortion versus mixer switch resistance

4.3 Circuit Realization

The complete circuit realization of the resistive QDAC TX is disclosed in this Section. Again, architecture validation and early-stage performance evaluation were performed using a comprehensive MATLAB model. Transistor-level simulations using SPECTRE were also used to validate the expected harmonic performance and out-of-band noise emission.

Both design and prototyping were realized using a 28 nm CMOS technology, with a supply voltage of 0.9 V.

4.3.1 RQDAC

An overview of the resistive QDAC architecture is shown in Fig. 4.24. The RQDAC is implemented as a 12-bit array, segmented into 7 bits binary and 5 bits unary in order to improve DNL and provide a monotone behavior. For the thermometric decoding, the same logic described in Sect. 3.3.1.3 was used in this implementation. Again, both binary and unary cells are derived from a parallel combination of the same unit cell.

Each one of the unit cells comprises a fixed unit resistance, which can be alternately connected to V_{DD} or G_{ND} through *SWH* and *SWL*, respectively. The number of active elements is given by the digital input word. The selected cells

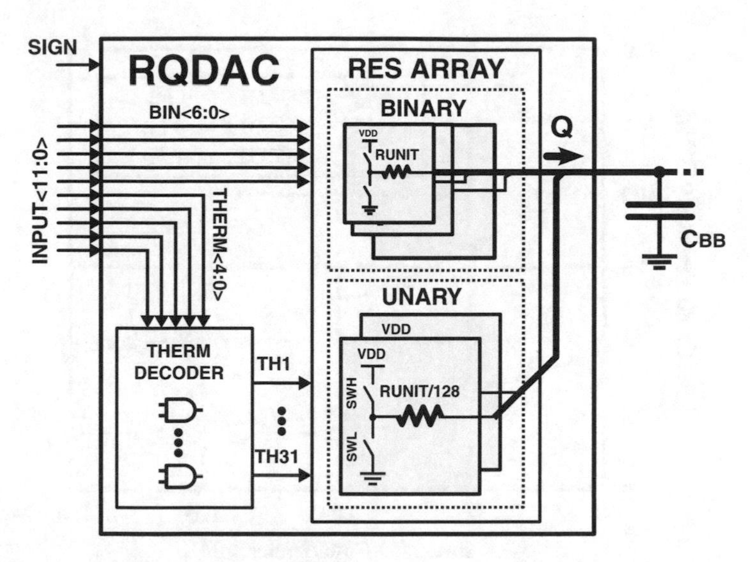

Fig. 4.24 RQDAC architecture

are combined in parallel, so that the RQDAC provides a linearly increasing conductance, proportional to *INPUT[0:11]*.

In the previous case of a capacitive QDAC TX, quantization noise could be scaled with the ratio between two capacitances, and the noise cutoff frequency was not affected by the RF load. The gate capacitance of the PPA input transistor was not sufficiently large to impact the baseband node's time-constant, allowing both baseband and unit capacitance to be reduced to a minimum while keeping all the noise filtering capabilities.

On the direct-launch transmitter, on the other hand, the low impedance RF load has a direct impact on the noise filtering, increasing the baseband conductance and shifting the noise f_{-3dB} to higher frequencies. The only way to reduce the noise cutoff frequency in this case (and thus benefit from intrinsic noise filtering) is by increasing the baseband capacitance accordingly. Figure 4.25 shows the equivalent noise cutoff frequency as a function of the baseband capacitance for various backoff conditions from 7 dBm maximum output power. According to Fig. 4.25, at least 125 pF baseband capacitance is required to guarantee noise filtering at the RX band (45 MHz offset) even at maximum output power. For a typical PAPR of 7 dB, the same 125 pF C_{BB} yields a noise cutoff frequency of roughly 15 MHz.

Once the baseband capacitance is defined, quantization noise becomes mostly a function of the unit resistance as demonstrated in Eq. (4.16). The R_{UNIT} value used in this implementation is chosen by increasing the unit resistance until the average quantization noise spectral density at 45 MHz offset (10 MHz integration window, from a 2 GHz modulated carrier) was lower than −165 dBc/Hz. Figure 4.26 shows the simulated noise spectral density versus the unit resistance, and the required

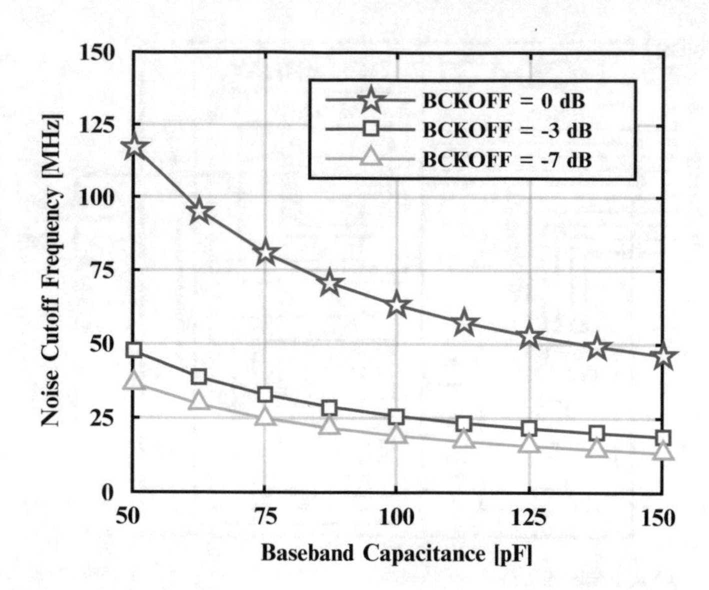

Fig. 4.25 Equivalent noise cutoff frequency as a function of the baseband capacitance

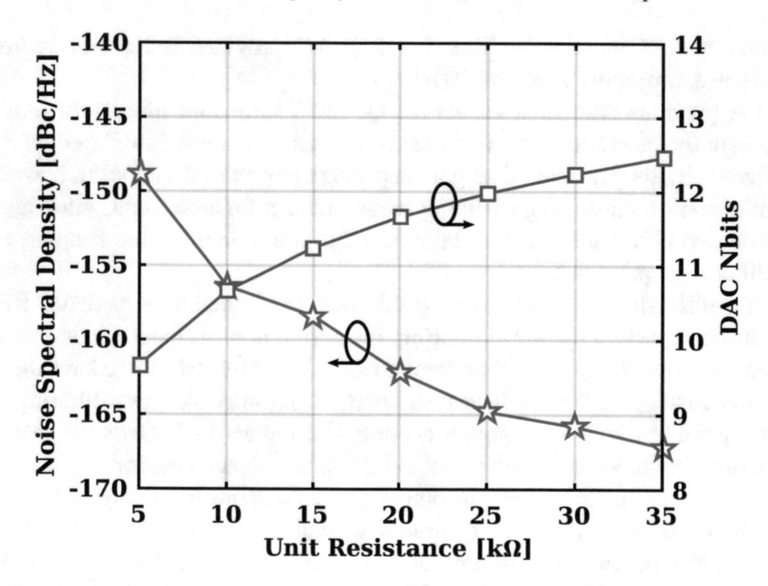

Fig. 4.26 Simulated noise spectral density versus the R_{UNIT}, and the required number of DAC elements for a 20 MHz bandwidth with a peak output power of 7 dBm

number of DAC elements (in bits) to provide up to 20 MHz bandwidth with a peak output power of 7 dBm. For this realization a unit resistance of 25 kΩ was chosen, thus requiring a 12-bit implementation.

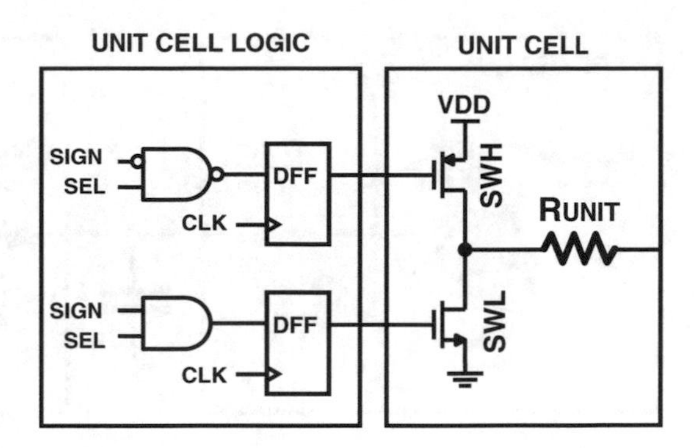

Fig. 4.27 RQDAC unit cell

4.3.1.1 Unit Cell

The simplified schematic of the resistive QDAC unit cell and the corresponding control logic are shown in Fig. 4.27.

Compared to the previous CQDAC, the resistive unit cell has a much simpler construction with only two switches that are operated at the much lower baseband sampling speed. Named *SWH* and *SWL* respectively, the V_{DD} and G_{ND} switches are selected according to the decoded selection bit SEL and the corresponding SIGN control bit. For obvious reasons, these two switches are never selected at the same time to avoid a "short" current between the supplies, which would also corrupt the amount of charge stored in the baseband capacitance.

As discussed in Sect. 4.2.4.1, the control switches *SWH* and *SWL* have a direct impact on the resistive DAC dynamic performance. The non-constant voltage-dependent switch ON-conductance degrades the system linearity introducing non-linear distortion. For this reason the sizes of both *SWH* and *SWL* are typically increased, so that the impact of the non-linear switch conductance in the overall unit cell conductance is minimized. At the same time, the OFF conductance should be reduced to minimize the charge leakage while the switch is not selected, otherwise the charge balance in the system can be affected. In this case, the R_{OFF}/R_{ON} ratio is maximized by using regular Vth transistors.

On top of these two contributors, another non-linear distortion mechanism should be accounted in the proposed implementation. As depicted in Fig. 4.28, the drain capacitances of both *SWH* and *SWL* switches combined with the contact parasitics of R_{UNIT} produce an additional load capacitance "C_P", that multiplied by the number of OFF-state unit cells at a given time, also appears as part of the load capacitance. Since the number of OFF cells is code dependent ($\#_{OFF} = \#_{TOTAL} - \#_{ON}$), this parasitic loading introduces signal distortion that affects more intensely the lower codes (when $\#_{OFF}$ is maximum) and it can hardly be pre-compensated.

The approach used here to minimize the dynamic performance degradation caused by "C_P" was to optimize the sizes of *SWH* and *SWL* and reduce the number of DAC elements though the series/parallel combination of unit cells. The whole DAC is implemented using 3.125 kΩ (25 kΩ/8) resistors, which are combined in

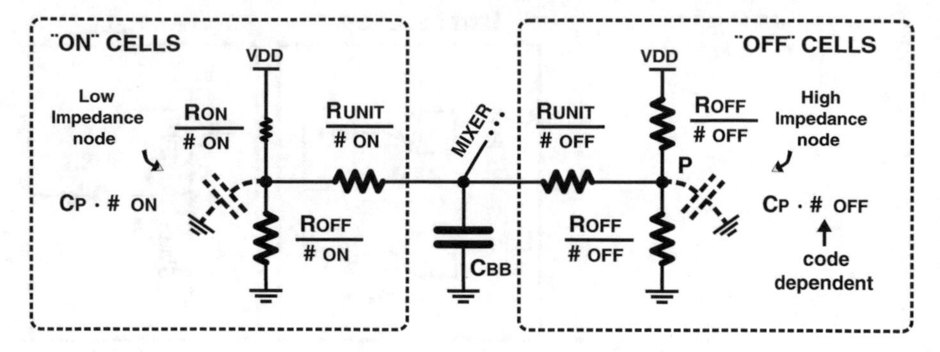

Fig. 4.28 Non-linear distortion mechanism due to OFF cells parasitic loading

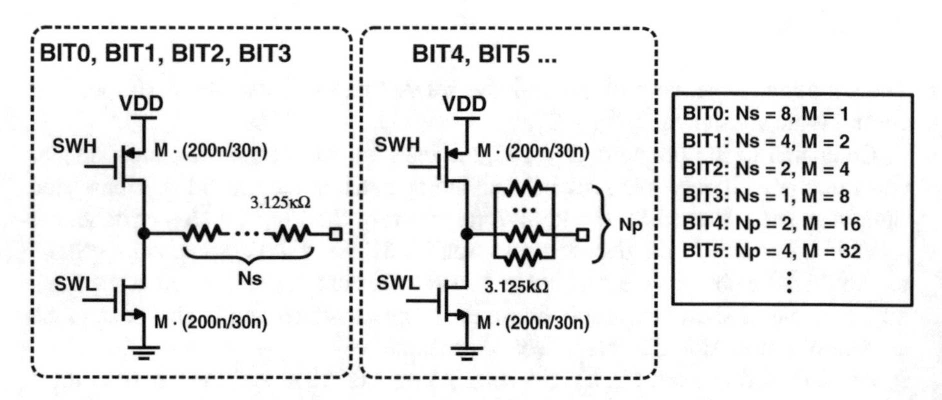

Fig. 4.29 RQDAC series/parallel bit construction

series in the 3 least significant bits (Fig. 4.29), and in parallel at the remainder. The total number of elements using this approach drops from 4095 to 525, minimizing significantly the parasitic capacitance at node *P*. Poly-type integrated resistors were also used for its lower parasitic capacitance and matching performance according to PDK technology data.

The impact of reducing the number of DAC elements (and hence the parasitic loading at node *P*) is demonstrated in Fig. 4.30 with an example spectrum using both DAC implementations. Figure 4.31 shows the expected third and fifth-order harmonics for various baseband frequencies using the proposed approach.

Notably, the main drawback of the proposed DAC topology is the matching degradation caused by the given LSB construction. Again, through layout extraction and extensive Monte Carlo simulations an expected yield of more than 80 % could be estimated, potentially improved by increasing the area taken by the unit resistors and selection switches.

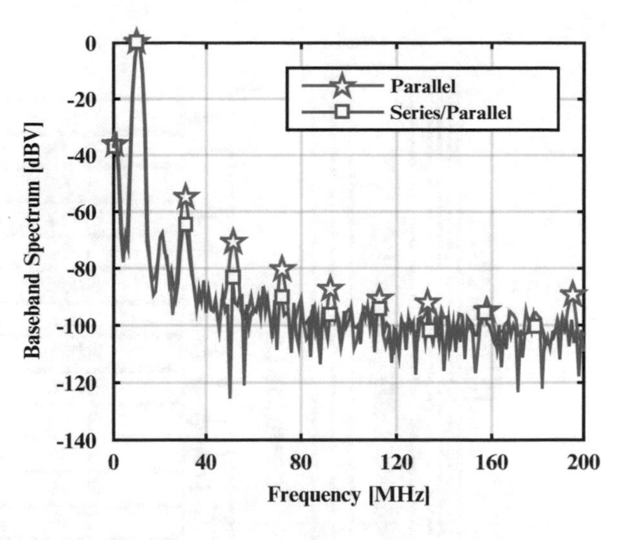

Fig. 4.30 Example spectrum showing the impact of reducing the code-dependent parasitic loading at node *P* through series/parallel resistor combination

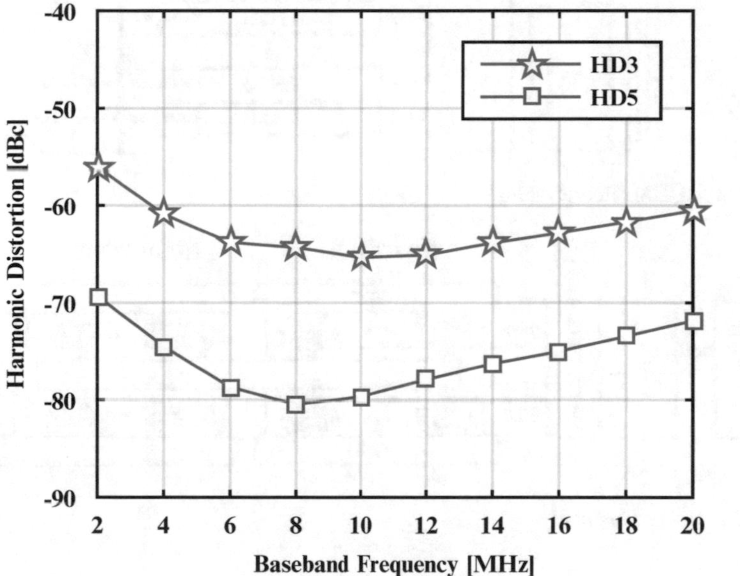

Fig. 4.31 Harmonic distortion of proposed RQDAC construction for various baseband frequencies

4.3.1.2 Layout

An overview of the RQDAC floorplanning is shown in Fig. 4.32. The 12-bit DAC array is divided into 38 lines, again with each line being occupied by a single binary or unary bit. Starting from the vertical central line, the multiple bit lines are spread both up and down, starting from the binary cells and followed by the unary. The unary cells are grouped according to the their thermometer logic functions (if looked

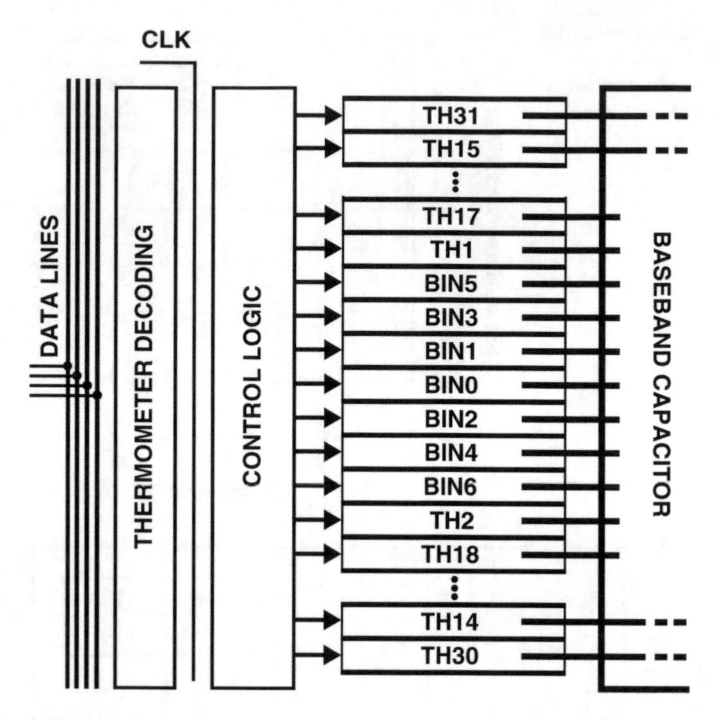

Fig. 4.32 RQDAC floorplanning

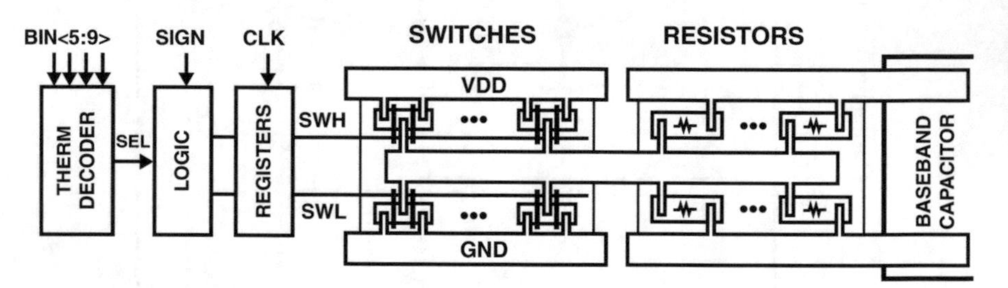

Fig. 4.33 Bit-line in detail

closely, the difference between TH1 and TH17 is only one logic gate, where the MSB is combined).

Different from the previous implementation, the RQDAC is fed from a single (left) side. The input data is first decoded and combined with the SIGN bit. Before connecting to the control switches, the control lines are re-timed using flip-flops to avoid errors due to switch-time skewing. The control signals are fed horizontally across the bit line (Fig. 4.33), and dummy switches are used to balance the capacitive loading seen from every control logic driver. Dummy cells are also widely applied to avoid border effects.

The baseband 125 pF MOM capacitor is placed on the right side of the array, and occupies a total area of 110 μm × 370 μm.

4.3.2 Mixer Design

In the direct-launch transmitter, the mixer implementation is critical. Since all the RF power is driven through the mixer switches, achieving the stringent linearity requirements with increased output power levels is challenging. In fact, once the PPA is removed from the signal path, the mixer switches become the primary signal distortion contributor in the charge-based transmitter.

As pointed in Sect. 4.2.4.2, the voltage dependence of the switch ON-conductance is the dominant distortion mechanism implied by the mixer, which becomes even more pronounced when low impedance RF loads are used. Achieving sufficient settling (as in the CQDAC TX) is not the main concern anymore, but rather how much the switch conductance changes across the baseband dynamic range.

The straightforward way to improve linearity in this case is to increase the switch conductance (decrease the switch resistance) until the impact of its voltage dependence in the total RF impedance ($R_{LOAD} + 2R_{MIXER}$) becomes negligible. However, as indicated by Fig. 4.34, in order to achieve a third-order counter intermodulation (CIM3) lower than −60 dBc, the mixer switch resistance would have to be at least 100 times lower than the RF load (using complementary switches), leading to an incredibly low switch resistance of 0.5 Ω. Providing such a large conductance with a single-type NMOS or PMOS switch would be simply impractical, and even with a complementary switch implementation the required sizes ($W_{SW} > 2$ mm, L_{MIN}) would also lead to excessive power consumption and LO feedthrough.

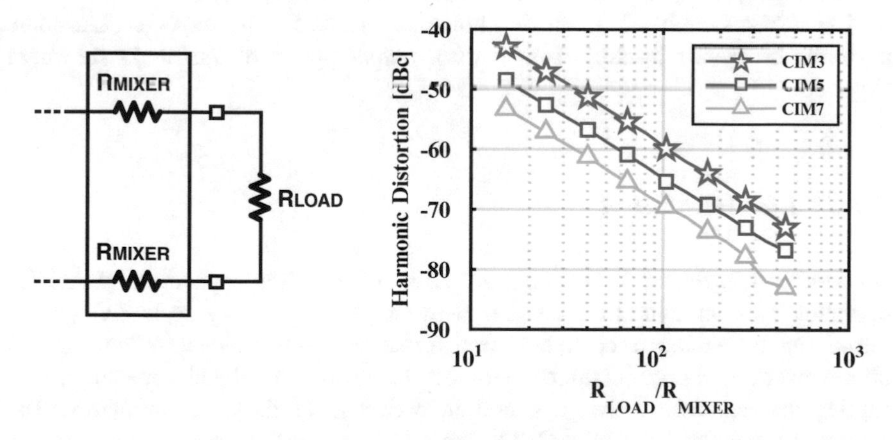

Fig. 4.34 Required switch resistance for a given mixer dynamic performance

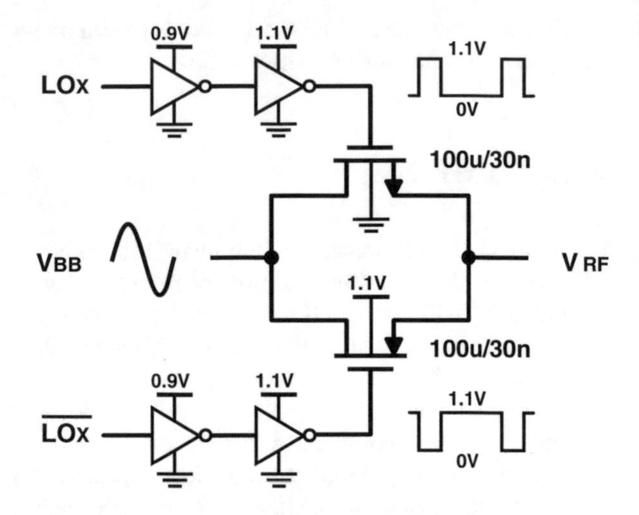

Fig. 4.35 Mixer switch schematic

A better compromise could be achieved in this design by keeping the switch sizes as small as possible, and boosting the ON-conductance by increasing the switch overdrive. A similar technique was used in the previous TX implementation by decoupling the LO signal and shifting the DC voltage with an external bias (Sect. 3.3.2.2). However, the non-zero OFF voltage created in this case would also increase the switch OFF-conductance, and the improvements provided by the larger overdrive would be masked by excessive leakage. The solution found in this implementation was to increase (by design) the supply voltage of the last LO driving stage from 0.9 to 1.1 V, so that the ON-conductance is increased without changing the OFF-conductance. Evidently, a valid concern about reliability is raised from the proposed solution, however the reduced duty cycle (25 %) is expected to compensate for the 22 % increase in the gate driving voltage, since short AC is believed to be less impacting than DC stress to oxide breakdown [Abo99].

The mixer switch schematic is shown in Fig. 4.35. The expected harmonic performance versus backoff from a peak output power of 7 dBm is shown in Fig. 4.36.

4.3.3 LO Generation

As described in Sect. 3.4, an important issue affecting the capacitive QDAC TX operation was an undesired overlap between the 25 % duty-cycle LO phases introduced by an increased hybrid transformer phase imbalance. When the LO phases overlap, charge is shared between the various baseband capacitors, corrupting the amount of charge stored in each one of these accumulators. The characterization of the CQDAC TX prototype was still possible thanks to the inclusion of a pulse generator reducing the duty-cycle at the $2 \cdot f_{LO}$ input, but the LO frequency was limited to roughly 1 GHz. To prevent the same issue from happening

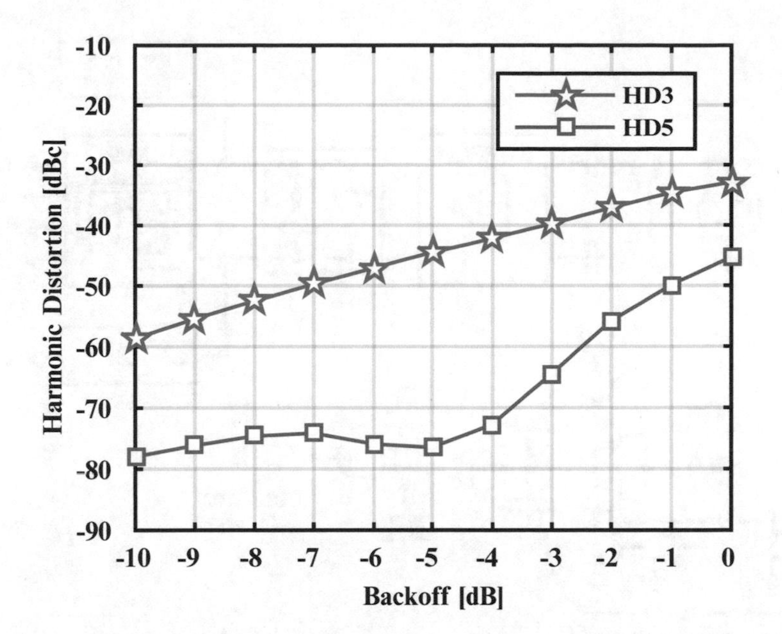

Fig. 4.36 Simulated harmonic performance for given mixer design

another time, a new method to create lower than 25 % duty-cycle LO phases is proposed.

Depicted in Fig. 4.37, first the LO signal is decoupled and buffered by the input stage. Second, a configurable-strength back-to-back inverter is used to restore the 180° phase difference between the differential LO components. The inclusion of the back-to-back inverters was crucial to fix the phase overlap, and a controllable-delay block was also added to reduce the duty cycle. Detailed in Fig. 4.37, by changing the driving strength of the falling edge in a specific inverter with a fixed load capacitance, it is possible to produce a configurable duty cycle LO signal. The LO generation block is able to provide duty cycles from 20 to 25 %, with worst-case loaded phase noise performance better than −160 dBc/Hz at 80 MHz offset (−159.2 dBc/Hz @ 40 MHz) for the minimum duty cycle configuration. Also, for a duty cycle larger than 24 %, the spectral noise density at 40 MHz drops to −167 dBc/Hz (Fig. 4.38).

4.3.4 Top-Level Description

A top-level diagram of the resistive QDAC TX is shown in Fig. 4.39. In addition to the above described RQDACs, mixer and LO generation, different blocks such as NOC, memory and clock dividers were also included.

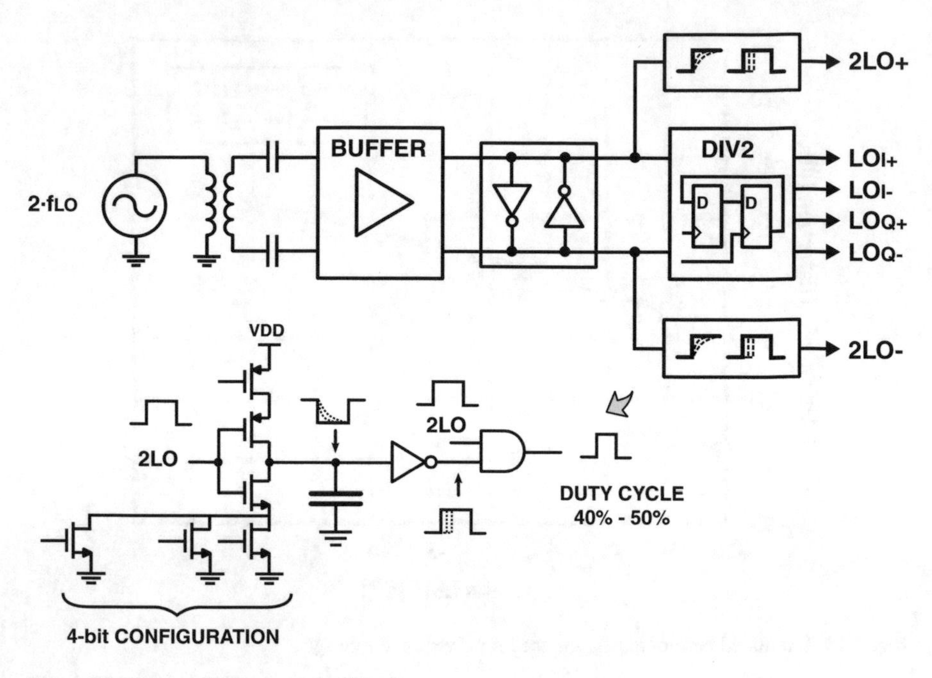

Fig. 4.37 Proposed LO generation block

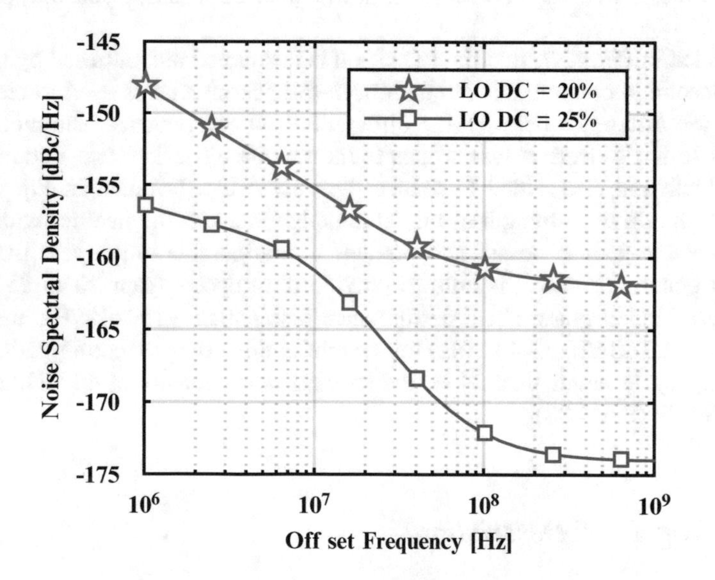

Fig. 4.38 LO Phase Noise performance at minimum and maximum duty cycle configurations

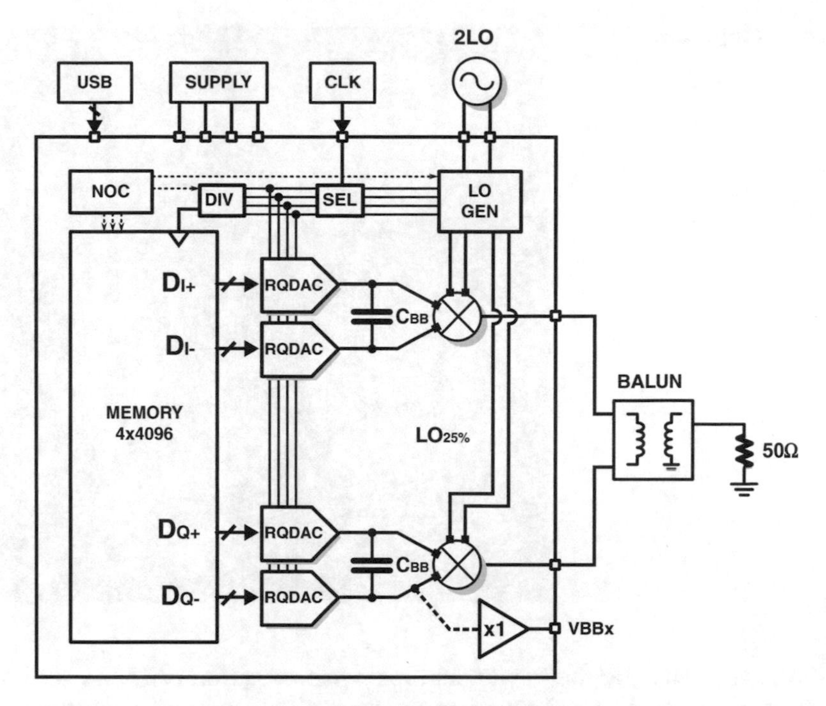

Fig. 4.39 Top-level block diagram of the RQDAC TX prototype

Since the resistive DAC architecture allows the baseband sampling frequency to be completely independent from the LO, an additional clock input with internal 50 Ω termination and buffering was also included. The memory clock input can be thus selected as a fraction of the LO frequency or the external clock (from 1/1 to 1/8), set by a sequence of frequency dividers that can be bypassed if needed.

The memory content is evaluated externally, using mathematical tool MATLAB, and later loaded into the integrated memory via SPI interface. Once loaded, the data is continuously cycled, first being de-interleaved and re-timed before reaching the RQDAC. Based on a wanted transmit signal, the algorithm evaluates the necessary amount of charge according to the Equations described in Sect. 4.2.1.

To address the memory size limitation faced in the previous TX realization, a RAM-cell-based 8k memory was used. Each memory position is 32-bit wide, designed to provide a 16-bit resolution to both I and Q data. Again, since the differential I and Q data can be distinct at any particular time ($D_{I+}[k] \neq D_{I-}[k]$), the transmit data is interleaved before loading to the memory, reducing the available number of memory positions by half. Nevertheless, with eight times the previously available memory size, EVM performance could finally be measured with this prototype.

Again, an additional test mode where the mixer switches can be independently switched OFF was also implemented. The baseband node observation without the influence of the mixer and RF load can provide valuable information about the

Fig. 4.40 Chip micrograph

RQDAC parameters, as well as its intrinsic dynamic performance. As such, the baseband nodes could be observed externally via four unity-gain amplifiers, as described in Sect. 3.3.4.

Finally, using a 28 nm CMOS technology, the chip occupies $1.05 \times 1.15\,\text{mm}^2$ (with pads), with an active area (including the entire transmitter except integrated memory and NOC) of $0.22\,\text{mm}^2$. Available spaces were filled with supply decoupling caps, which again are not believed to be decisive on achieving the reported results. A chip micrograph is provided in Fig. 4.40.

4.4 Measurement Results

4.4.1 Measurement Setup

For the RQDAC TX measurements, the prototype chip is directly bonded to a custom-made PCB (FR4—4 layers) using conventional wire-bonds. Distinct supplies are assigned to each one of the sensitive blocks, including the analog RQDAC supply and the mixer. These supplies are separated from the other switching intensive blocks, such as the memory and clock buffering cells. The baseband voltage buffers are also supplied from a dedicated 1.8 V source. In total, five different supplies are used (3×0.9 V, 1×1.1 V and 1×1.8 V), all provided by Agilent N6705 regulated supplies with multiple outputs.

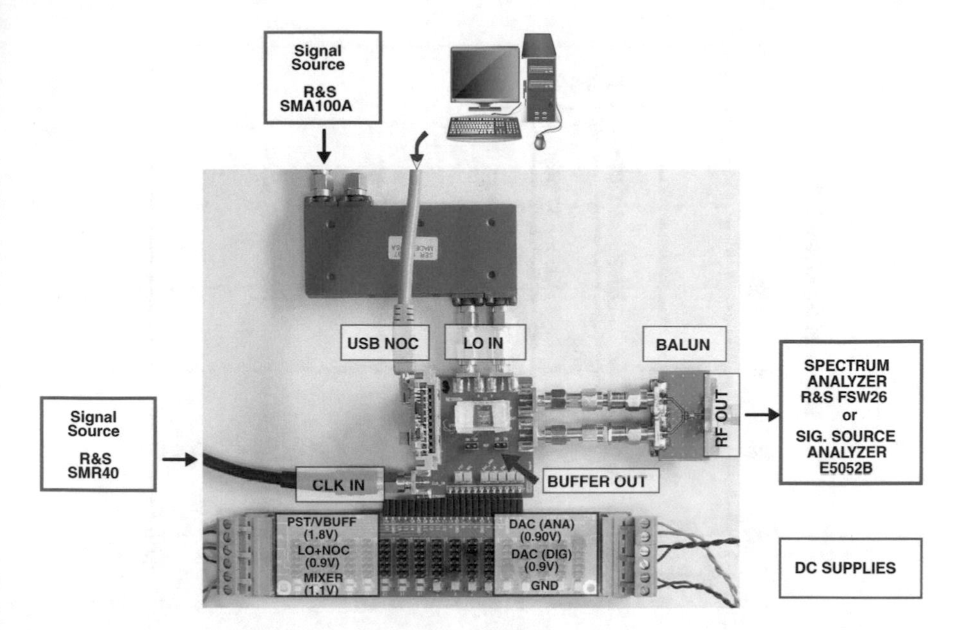

Fig. 4.41 RQDAC TX measurement setup

The RF output was directly measured using the R and S FSW-26 spectrum analyzer, while the baseband voltages were buffered with a differential active probe (TEKTRONIX 1163). Without any output amplifying stage or bias tee, the mixer output directly drives the RF load. The differential to single-ended conversion is done using two different external transformer-based baluns (MURATA) for the frequencies of 900 MHz and 2.4 GHz, AC coupled to the mixer output. The 50 Ω RF load is given by the measurement equipment input impedance.

The LO differential signal is input from a KRYTAR hybrid coupler, sourced from a R and S SMA100A. The clock input signal is provided from a R and S SMR40.

An overview of the measurement setup is shown in Fig. 4.41.

4.4.2 RQDAC Measurement Results

As with the previous charge-based design, the TX measurements starts from the RQDAC characterization. By switching OFF the mixer switches, the RQDAC output could be observed externally without the influence of the RF load. For the RQDAC parameter extraction, implementation aspects prevented the estimation of the baseband capacitance through the measurement of the input current, as done in the previous TX. Instead, the RQDAC calibration was done by fixing a baseband capacitance of 252 pF (estimated with layout extraction) and sweeping the unit resistance value until best harmonic performance was achieved.

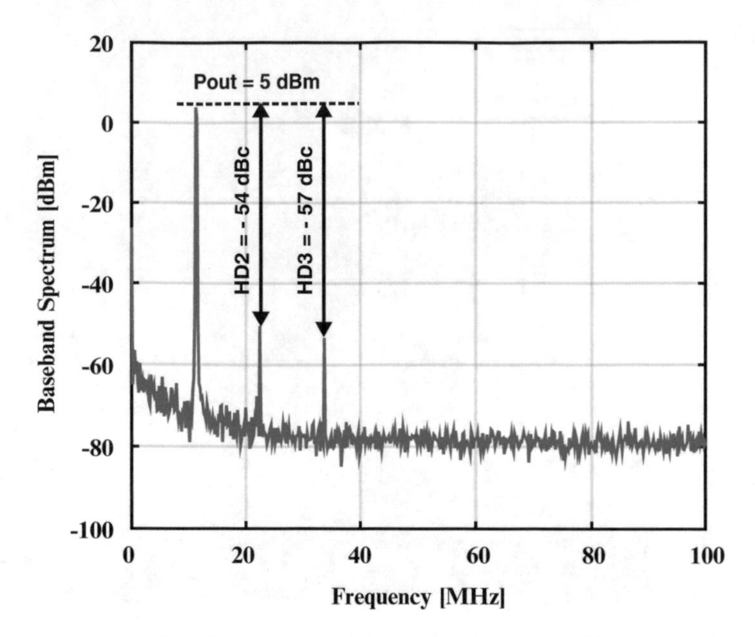

Fig. 4.42 Measured RQDAC baseband output spectrum with mixer OFF

Figure 4.42 shows the output spectrum of a 10 MHz baseband tone (500 MS/s) for the maximum 550 mVpp (single-ended) baseband swing when a unit resistance of 26.75 kΩ is considered.

The dominant harmonics (second, third and fifth) at both full swing and 3 dB backoff are shown Fig. 4.43. As noted, at 3 dB backoff every harmonic is lower than −55 dBc, demonstrating superior linearity as required by advanced wireless communication systems. Naturally, the second-order harmonic are further reduced in differential operation.

4.4.3 RQDAC TX Measurement Results

After fine tuning of the RQDAC unit resistance and baseband capacitance, the mixer was switched back ON and several different measurements were realized in order to assess the TX performance with respect to noise, harmonic distortion, spurious emission, EVM and power consumption. The transmitter performance was characterized at 900 MHz and 2.4 GHz, as an attempt to demonstrate its functionality at both low and high cellular bands, as well as WiFi/Bluetooth applications in the ISM bands.

After de-embedding losses from both cable and balun, a peak RF output power of 3.5 dBm was measured at both LO frequencies. Figure 4.44 shows an example spectrum of a 10 MHz single-tone (transmitted at 900 MHz with an output power

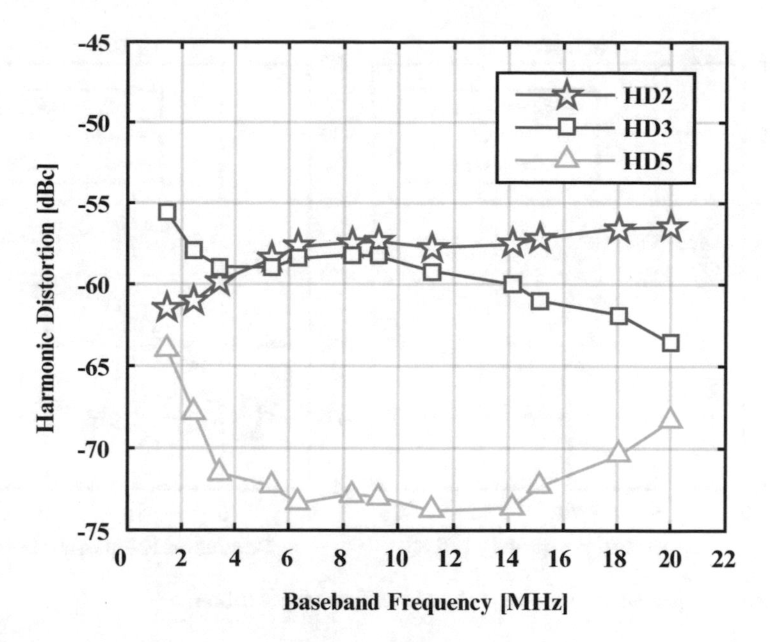

Fig. 4.43 Measured RQDAC harmonic performance with mixer OFF

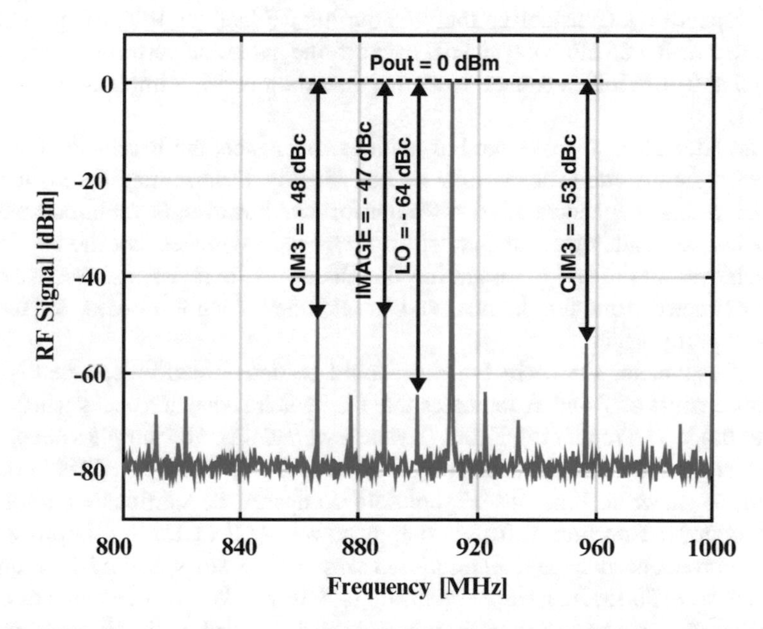

Fig. 4.44 Example RF output spectrum for a 10 MHz single-tone modulated at 900 MHz, with 0 dBm output power

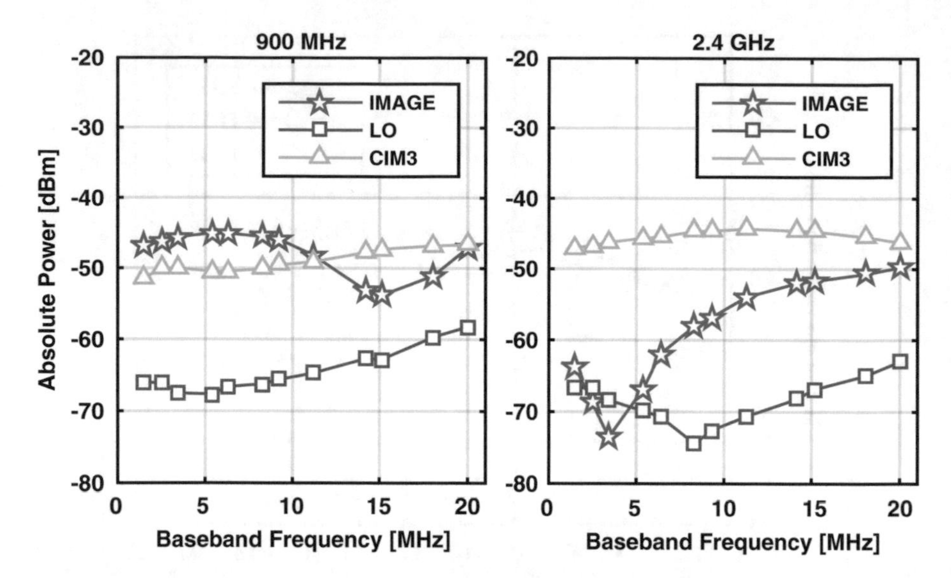

Fig. 4.45 RF spurious emission versus baseband frequency at 0 dBm

of 0 dBm), where a native LO and Image suppression of −64 and −48 dBc can be noted, respectively. Considering that without the RF load the RQDAC provides an HD3 lower than −55 dBc also at 3 dB backoff, the harmonic performance shown in Fig. 4.45 at 0 dBm indicates that excessive distortion is being implied by the mixer switches.

As an attempt to improve the harmonic performance, the baseband signal was tentatively pre-distorted for the mixer's non-ideality. Determining the switch conductance voltage dependence is not straightforward however. Several measurement approaches were attempted to characterize the mixer switches, but the best results were achieved in the end by assuming the characteristic resistance versus voltage profile extracted from simulations, and "stretching" it while looking for the best harmonic performance.

The X parameter shown in Fig. 4.46 could be determined independently since any combination of Y and Δ indicates that the switch resistance peaks when V_{BB} is equal to 0.6 V. The remaining Y and Δ values were defined by simple sweep.

With an X, Y and Δ of respectively 0.6, 3.9 and 2.5, CIM3 and CIM5 could be reduced, as shown in Fig. 4.47. Figure 4.48 compares the spurious emission with and without predistortion at 0 dBm output power. At 2.4 GHz, the improvements are more pronounced since the calibration was done at this specific LO frequency. Nonetheless, at 900 MHz improvements up to 5 dB in CIM3 can be noted as well.

The single-tone spurious emission was also verified with different backoff conditions and sampling frequencies (Fig. 4.49). At 500 MS/s, CIM3 as well as the higher frequency odd harmonics are always below −50 dBc, for any baseband frequency. LO feedthrough and Image rejection, in turn, are always lower than −40

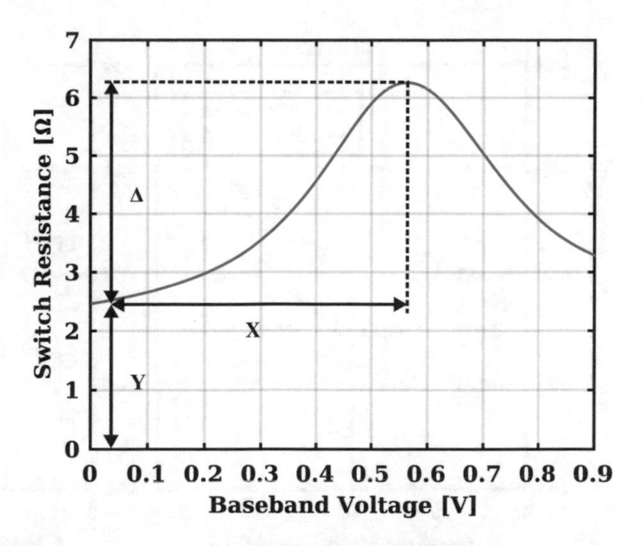

Fig. 4.46 Mixer switch resistance voltage dependence used to pre-distort the baseband voltage

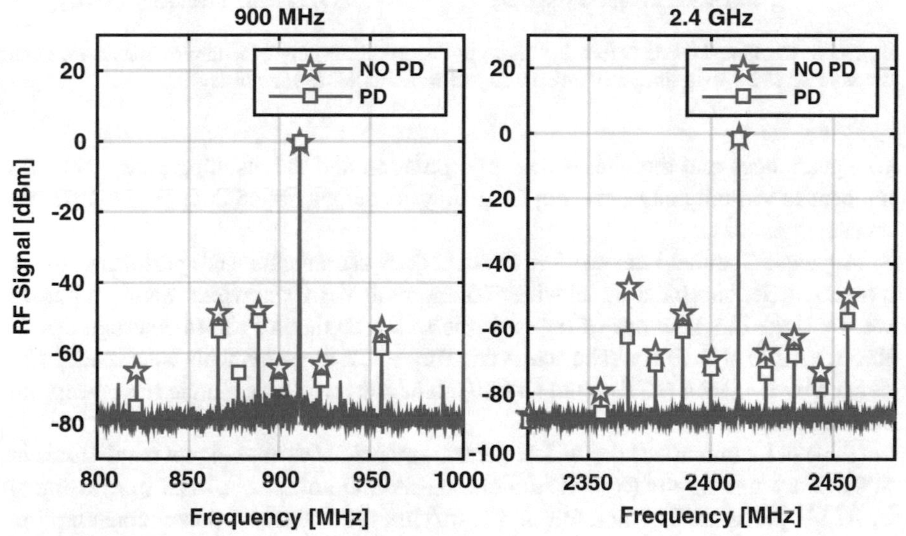

Fig. 4.47 RF spectrum before and after baseband pre-distortion

and −50 dBc respectively. Running the RQDAC at higher speed (1000 MS/s) did not bring much improvements in terms of harmonic performance as also noted in Fig. 4.49, reason why the lower 500 MS/s sampling frequency was preferred during ACLR measurements.

The measured sampling aliases are shown in Fig. 4.50 for multiple baseband and sampling frequencies. As in the previous charge-based TX, the output spectrum is clearly shaped by a sinc^2 transfer function, which in worst case adds at least 20 dB of additional attenuation at 500 MS/s. The continuous-time RC charging of

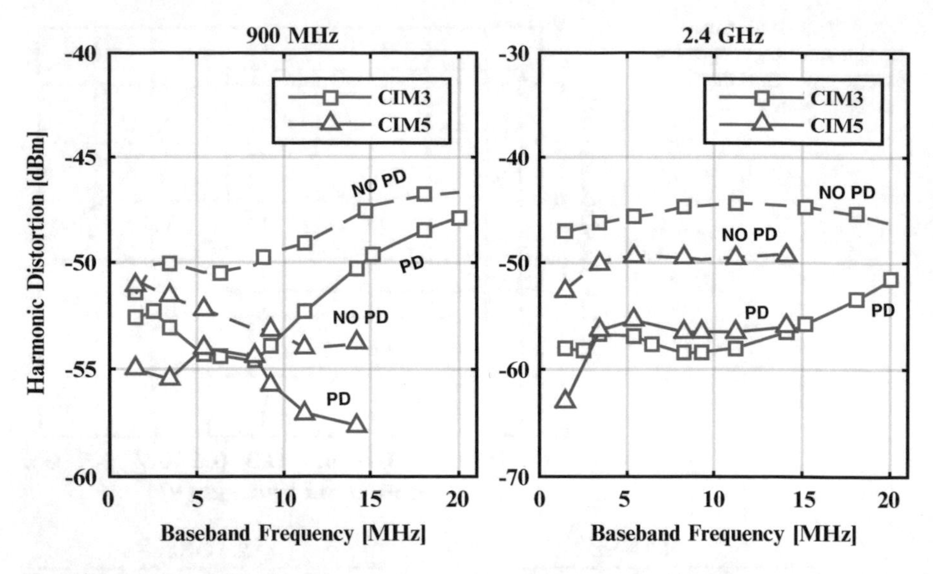

Fig. 4.48 CIM3 and CIM5 before and after pre-distortion. As noted, the improvements are more pronounced at 2.4 GHz since calibration was performed at this LO frequency

C_{BB} guarantees that the quasi-linear interpolation and the resulting alias reduction happens at virtually any sampling frequency, as demonstrated through the different measurement cases.

Adjacent Channel Leakage Ratio (ACLR) measurements were performed using a multi-carrier baseband signal with maximum 20 MHz bandwidth. Random phases were assigned to each one of the multiple tones, so that a Peak-to-Average Power Ratio (PAPR) of 7 dB could be achieved. The ACLR measurements were realized as defined by E-UTRA (LTE10 and LTE20) standards, with an example screen capture shown in Fig. 4.51.

Table 4.1 summarizes the ACLR measurements for various signal bandwidths at 500 MS/s. Running the RQDAC a higher speed did not bring a clear improvement in ACLR performance, therefore not justifying the additional power consumption that would be required in this case. The baseband pre-distortion, on the other hand, provided an average 2 dB improvement to both ACLR 1/2.

For the EVM measurements, different WLAN-like transmit signals were generated using QPSK, 16QAM and 64QAM. The constellation plots for each one of these cases are shown in Fig. 4.52, with their respective measured EVM shown in Table 4.2. As noted, outstanding EVM performance below 1.64 % could be achieved at both LO frequencies.

The out-of-band noise performance of the resistive QDAC TX at both 900 MHz and 2.4 GHz are respectively shown in Figs. 4.53 and 4.54, using in both cases a 10 MHz single-tone (500 MS/s) modulated carrier, so that quantization noise is also included. At 45 MHz offset, the measured noise spectral density at both LO

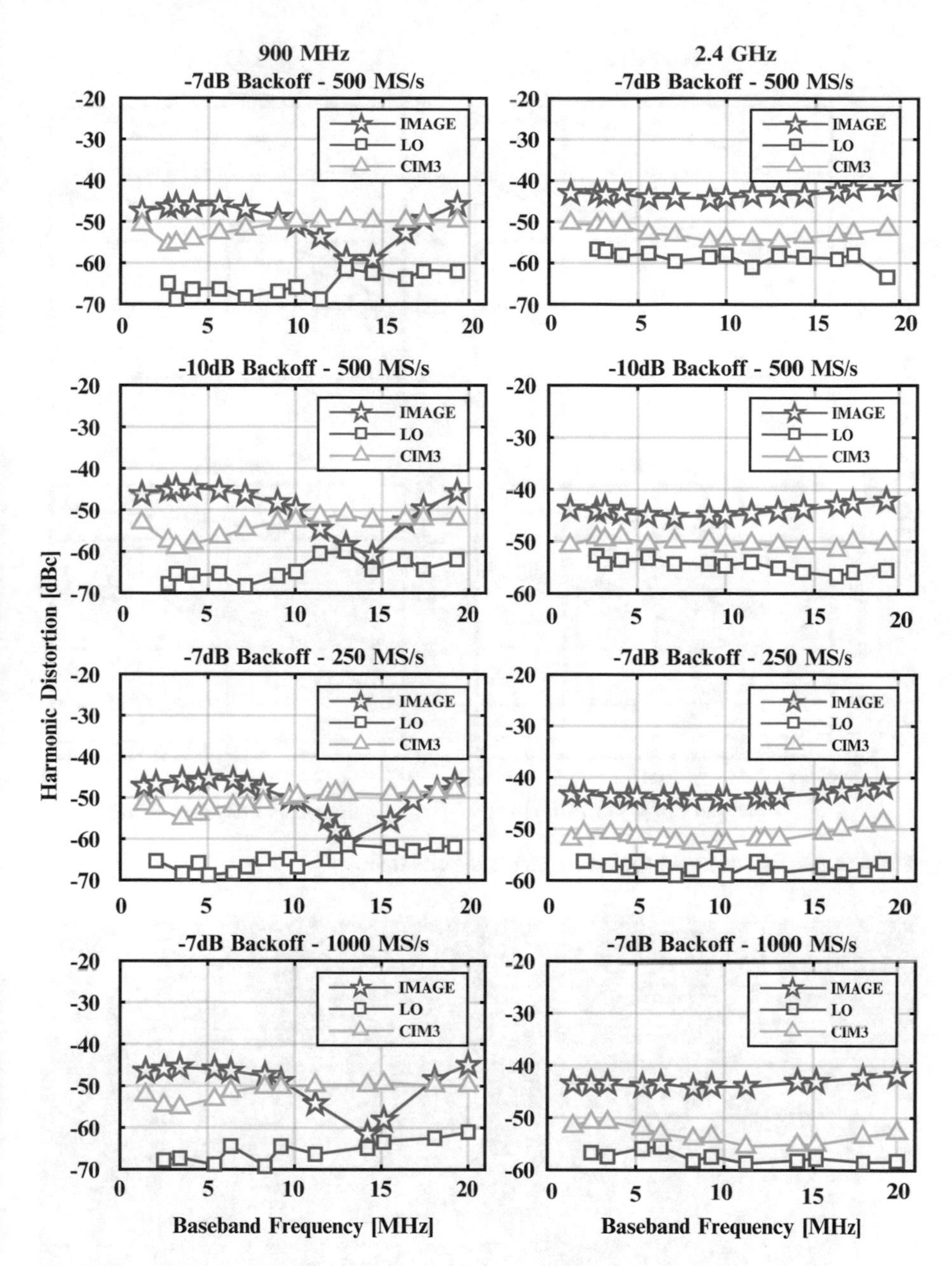

Fig. 4.49 Single-tone spurious emission for various backoff values and sampling frequencies

frequencies are notably below −159 dBc. Moreover, thanks to the intrinsic noise filtering capabilities, at 7 dB backoff the noise performance drops by less than 2 dB as shown in Fig. 4.55.

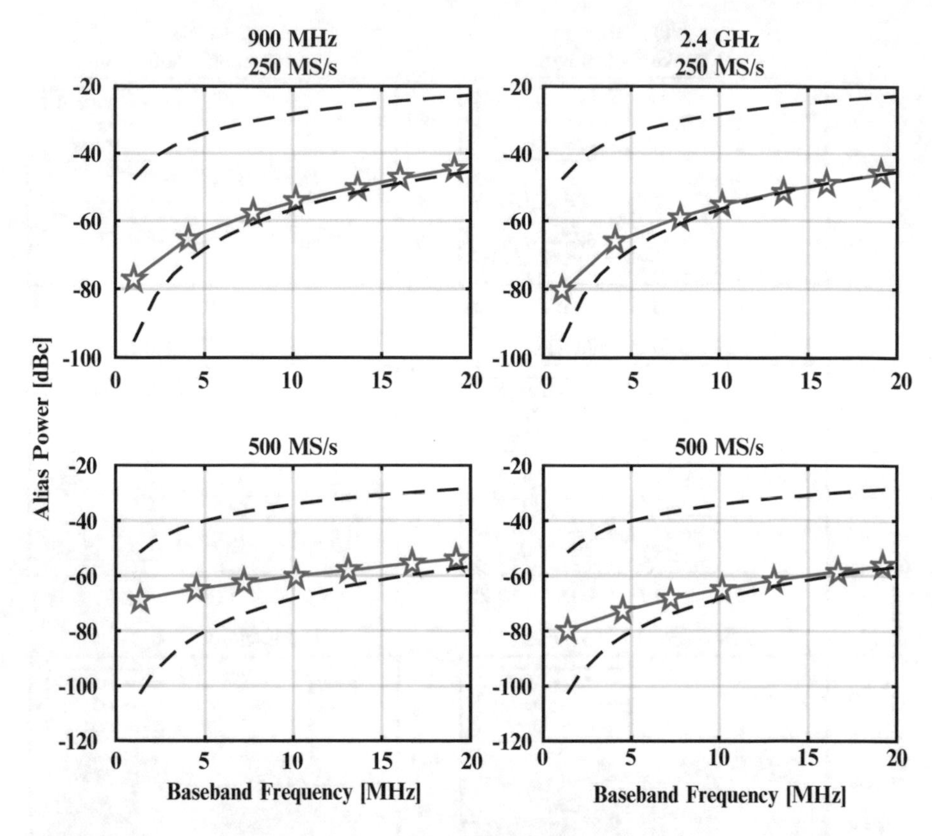

Fig. 4.50 Measured sampling aliases at both 250 and 500 MS/s

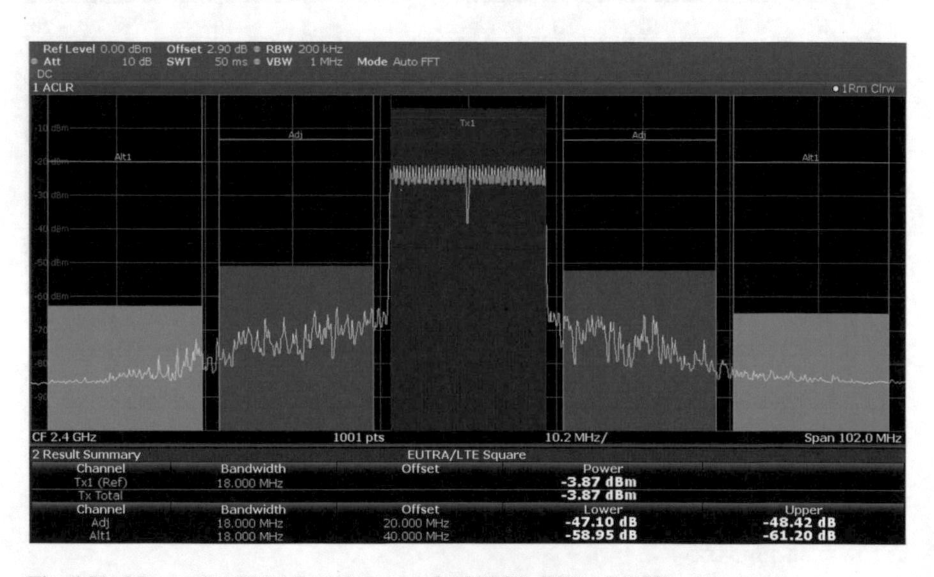

Fig. 4.51 Measured ACLR1/2 performance for 20 MHz BW at 2.4 GHz

Table 4.1 ACLR1/2 performance

ACLR1/2 performance				
BW		5 MHz	10 MHz	20 MHz
ACLR1/2 @ 900 MHz	[dB]	−48.8/−62.8	−48.3/−61.4	−49.2/−59.5
ACLR1/2 @ 2.4 GHz	[dB]	−46.8/−58.9	−47.2/−58.5	−47.1/−58.9

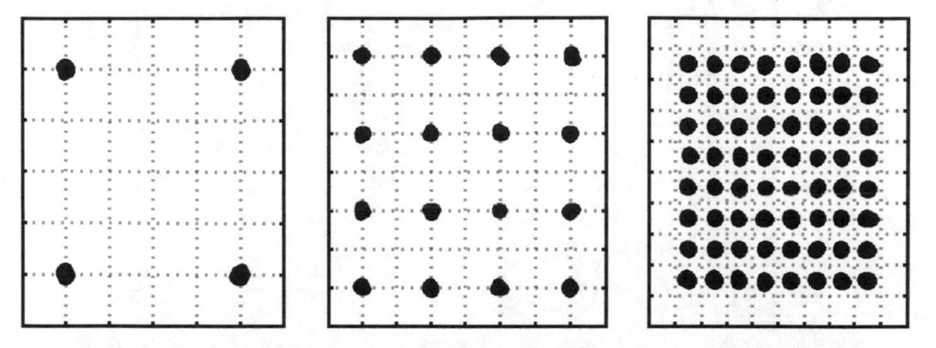

Fig. 4.52 Measured QPSK, 16QAM and 64QAM constellation plots

Table 4.2 EVM performance

EVM performance				
Modulation		QPSK	16QAM	64QAM
EVM @ 900 MHz	[dB]	−35.7	−36.9	−36.1
EVM @ 2.4 GHz	[dB]	−35.8	−37.1	−37.0

Finally, a performance summary (Tables 4.3 and 4.4) and comparison table are provided (Table 4.5). Driven at baseband sampling rate the power consumed in the resistive QDAC digital operation (decode, register and switching) is reduced by a factor of 7 when compared to [Par15b], providing an even simpler and more power efficient solution for charge-based architectures. Leveraged by the incremental-charge-based operation, the entire signal path comprising RQDAC, mixer and RF load achieves an improved efficiency of 5.8 % at 7 dB backoff, only degraded by the LO power consumption—over designed due to the external LO utilization in the test chip. In a transmitter with integrated VCO, the LO chain power consumption is expected to be reduced by approximately 60 %.

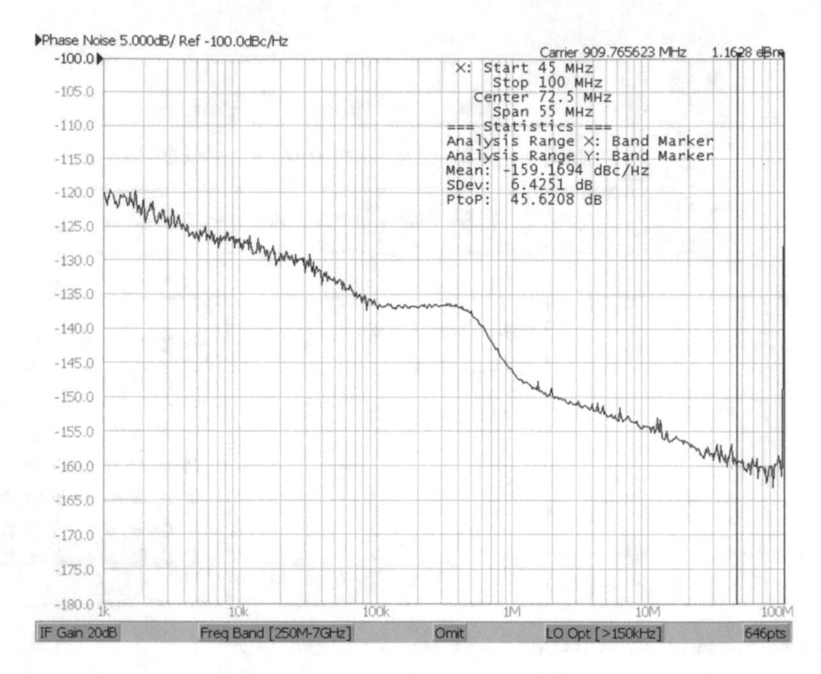

Fig. 4.53 Measured out-of-band noise at maximum output power for 900 MHz modulated carrier (10 MHz single-tone sampled at 500 MS/s)

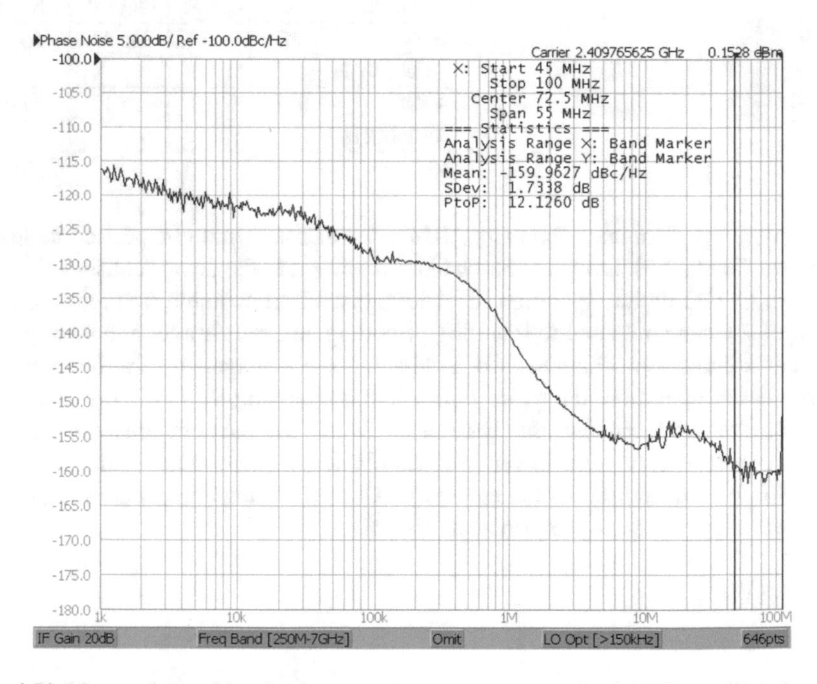

Fig. 4.54 Measured out-of-band noise at maximum output power for 2.4 GHz modulated carrier (10 MHz single-tone sampled at 500 MS/s)

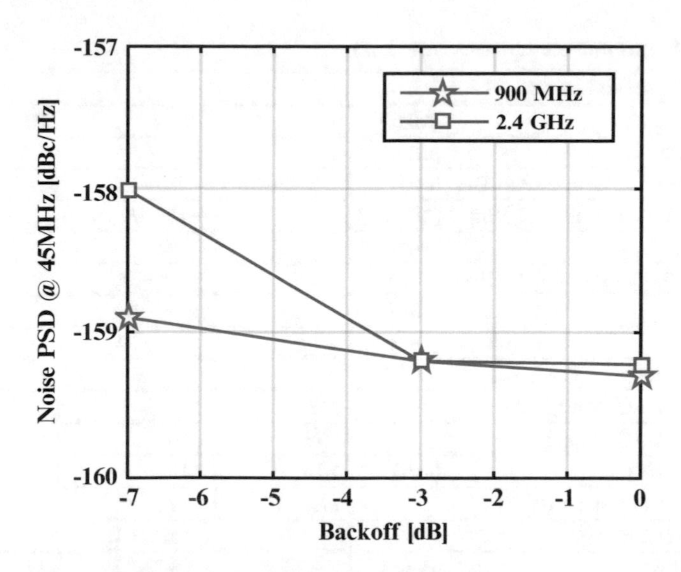

Fig. 4.55 Measured out-of-band noise at different backoff values

Table 4.3 Performance summary (900 MHz)

Performance summary @ LO = 900 MHz				
RF bandwidth	[MHz]	5	10	20
Maximum pout	[dBm]	3.5	3.5	3.5
ACLR1/ACLR2 @ −3.8 dBm	[dB]	−48.8/−62.8	−48.3/−61.4	−49.2/−59.5
Noise @ 3.5 dBm/−3.5 dBm	[dBc/Hz]		−159.2/−158.9	
Offset	[MHz]	45		
LO Feedthrough @ −3.5 dBm	[dBc]	<−60		
Image @−3.5 dBm	[d Be]	<−45		
CIM3 @−3.5 dBm	[dBc]	<−50		
Consumption @ −3.5 dBm				
DAC (charge intake)		2.36	2.45	2.99
DAC (digital)		1.35	1.44	1.57
Mixer	[mA]	1.59	1.59	1.59
LO generation		5.75	5.57	5.75
Supply voltage	[V]	0.9/1.1/1.8		
Active area	[mm^2]	0.22		
Process	[nm]	28 nm		

4.5 Conclusion

In this Chapter, the operating principles, circuit realization and measurement results of the resistive charge-based TX are disclosed.

Table 4.4 Performance summary (2.4 GHz)

Performance summary @ LO = 2.4 GHz				
RF bandwidth	[MHz]	5	10	20
Maximum pout	[dBm]	3.5	3.5	3.5
ACLR1/ACLR2 @ −3.8 dBm	[dB]	−46.8/−58.9	−48.0/−58.5	−47.1/−58.9
Noise @ 3.5 dBm/−3.5 dBm	[dBc/Hz]		−159.9/−158.0	
Offset	[MHz]		45	
LO Feedthrough @ −3.5 dBm	[dBc]		<−55	
Image @−3.5 dBm	[dBc]		<−40	
CIM3 @−3.5 dBm	[dBc]		<−50	
Consumption @−3.5 dBm				
DAC (charge intake)		2.20	2.28	2.81
DAC (digital)		1.72	1.80	1.93
Mixer	[mA]	4.19	4.19	4.19
LO generation		17.7	17.7	17.7
Supply voltage	[V]		0.9/1.1/1.8	
Active area	[mm^2]		0.22	
Process	[nm]		28 nm	

Based on the observation that the first chip's power consumption was highly impacted by the PPA bias current, a direct-launch implementation where the PA is directly driven using the QDAC was targeted. The QDAC was implemented using a 12-bit conductance array, which proved to be the most area efficient way of increasing the charge capacity of the QDAC. Instead of delivering packets of charge at LO rate as in the first implementation, now the required total charge is transferred to the baseband node by charging and discharging C_{BB} in continuous-time. Since the DAC switches are now operated at baseband speed, further improvements in power consumption could also be achieved when compared to the first realization.

All the noise filtering capabilities provided by the capacitive charge-based TX are also seen in its resistive implementation, with only one exception: the quantization noise in the RQDAC transmitter does not scale with the ratio between two capacitances, but it's rather dependent on the absolute value of the unit resistance and baseband capacitance. Intrinsic noise filtering (including quantization) is provided though, but in the direct-launch implementation the baseband node's conductance is increased significantly when a low impedance RF load is used. To keep the noise cutoff frequency below 20 MHz so that out-of-band noise emission is reduced, the baseband capacitance has to be increased accordingly.

Different from the CQDAC, the harmonic performance of the resistive charge-based TX does not depend on settling. In fact, once the PPA is removed from the signal path, the dominant contributor to signal distortion becomes the voltage dependence of the mixer switches, reduced in this implementation by using complementary switches and boosting their conductance with increased overdrive.

Table 4.5 Comparison table

REF		Codega et al. JSSC 2014		Eloranta et al. JSSC 2007	Alavi et al. TMTT2014	CQDAC	RQDAC	
Architecture		ANALOG		DDRM	RFDAC	QDAC	QDAC	
							900 MHz	2.4 GHz
RF bandwidth	[MHz]	10	20	5	20	20	20	
Max output power	[dBm]	6	6	NA	22.8	6[a]	3.5	
(RF)DAC	[bits]	10		10	13	10	12	
MaxBBClk	[MHz]	NA		307	300	128	500	
Noise @Offset Power	[dBc/Hz]	<−159[b] @45 M 2.8 dBm	<−158[b] @80 M 0dBm	−146 @190 M −2 dBm	−160[c] @200 M NA	−155 @45 M 1 dBm	−159.2/−158.9 @45 M 3.5 dBm/−3.5 dBm	−159.9/−158.0
ACLR/ACLR2	[dBc]	−43.4/−55 @2.8 dBm	−42.5/−55 @4 dBm	−58/−61 @−2 dBm	NA	−42/−47 @1dBm	−49.2/−59.5 @−3.8 dBm	−47.1/−58.9
CIM3	[dBc]	NA	−57.1 @2.3 dBm	NA	NA	<−50 @1 dBm	<−50 @−3.5 dBm	
Consumption		@2.8 dBm	@4 dBm	@NA	@22.8 dBm	@1 dBm, LO = 1 GHz	@−3.5 dBm	
Modulator						0.5/12.1	2.7/1.4[d]	2.5/1.7
LO	[mW]			92 @ ANA	33	5.7	5.1	15.9
MIXER				65 @ DIG		–	1.7	4.6
PPA					–	23	–	–
Total		97	98	157	NA	41.3	11.1	24.8
Supply	[V]	1.8		1.2	1.2	0.9/1.8	0.9/1.1/1.8	
Active area	[mm^2]	1.3			0.45	0.25	0.22[e]	
Process	[nm]	55 LP		130	40	28	28	

[a]Maximum output power limited by maximum DAC capacitance
[b]Using Baseband reconstruction filter
[c]Static measurement (does not include quantization noise)
[d]Modulator power consumption split into (charge intake)/(digital)
[e]Approx. 0.34 mm^2 if an integrated balun were to be considered

Thanks to the reduced switching speed, the design of the RQDAC is also more relaxed when compared to the capacitive counterpart. Further details about the prototype realization are given in Sect. 4.3.

The transmitter prototype was characterized at both 900 MHz and 2.4 GHz. With a peak output power of 3.5 dBm (from a 0.9 V supply), measurements with baseband frequencies ranging from 1 to 20 MHz show a maximum LO feedthrough and Image of −55 and −40 dBc respectively. At seven 7 dB backoff, CIM3 at any baseband frequency is always below −50 dBc at both LO frequencies, and as in the previous implementation the sampling aliases are shaped by a sinc^2 transfer function, which corresponds to at least 20 dB of additional attenuation if compared to a conventional architecture. The ACLR 1/2 performance for a 20 MHz bandwidth signal are better then −47 and −59 dB, with a measured EVM performance of 1.6 %. Finally, at 45 MHz offset the modulated noise power density below −159 dBc/Hz was measured at both LO frequencies, and what is very unique is the fact that even in backoff conditions, the noise performance is not significantly degraded, also thanks to the intrinsic noise filtering capabilities of the charge-based architecture.

To conclude, with the achieved out-of-band noise performance and an active core area consumption of only 0.22 mm^2, this architecture achieves what is—to the author's knowledge—the best out-of-band noise performance versus area consumption when compared to other similar works. ACLR and EVM performance are also among best.

Chapter 5
Conclusion

5.1 Summary

With the fast evolution of wireless systems, communication speed is constantly being taken to unprecedented levels. The concurrent expansion of system user capacity and peak data rates could only be achieved over the years with a corresponding increase in spectrum efficiency and larger bandwidths. Especially when the SAW filter is removed, the required transmitter frontend must perform remarkably both in terms of (out-of-band) noise and linearity, which is difficult to realize without sacrificing power and/or area consumption.

The analysis of current literature shows that with regard to CMOS RF transmitter implementations, the state-of-the-art is divided into analog and digital-intensive architectures. In terms of out-of-band noise, analog intensive architectures are undoubtedly the best performing implementations. However, their improved noise performance is typically achieved through extensive low-pass filtering, which has a significant impact in area consumption. Digital-intensive implementations, on the other hand, are by far the most portable, area efficient and scaling friendly. However, the lack of filtering (for both noise and aliases) makes it very challenging to meet the stringent out-of-band noise requirements in SAW-less operation.

To relax this trade-off, an incremental charge-based architecture is proposed. Through the combination of charge-domain operation with incremental signaling, this architecture aims to provide the best of both worlds, meaning the reduced area and high portability of digital-intensive architectures and an improved out-of-band noise performance given by intrinsic noise filtering capabilities. In a nutshell, the proposed architecture offers a significant reduction of the sampling aliases due to a quasi-linear interpolation provided inherently, and an improved out-of-band noise performance achieved thanks to an intrinsic single-order lowpass filter that comprises the signal path. Quantization noise is also reduced.

Two full-featured incremental charge-based TX implementations were disclosed in this book, differing on how the charge-based DAC (QDAC) is implemented, and

P.E. Paro Filho et al., *Charge-based CMOS Digital RF Transmitters*, Analog Circuits and Signal Processing, DOI 10.1007/978-3-319-45787-1_5

the RF load being driven: In the first realization, the RF load corresponds to the input capacitance of a PPA stage, and the QDAC is implemented with a controllable capacitance that is first pre-charged and then connected to a baseband capacitance. Fractions of the total charge required per sampling period are conveyed at LO speed in discrete packets, sized by adjusting the DAC capacitance accordingly. In the second implementation, the ability of delivering more power using the charge-based architecture was investigated. Based on the observation that the first chip's power consumption was highly impacted by the PPA bias current, a direct-launch implementation where the PA is directly driven using the QDAC was targeted. The benefits in this case not only include removing a power-hungry block from the signal path (PPA), but also increasing the effectiveness of pre-distortion by being able to directly control the PA input. The resistive QDAC implementation proved to be the most area efficient way of increasing the charge capacity of the QDAC. Instead of delivering packets of charge at LO rate as in the first implementation, the required total charge is transferred to the baseband node by charging and discharging the baseband capacitance in continuous-time, allowing the DAC switching speed to be chosen independently from the LO frequency. Moreover, since the DAC switches are operated at baseband speed (and not LO), further improvements in power consumption could also be achieved when compared to the first realization.

Both charge-based TX implementations were prototyped using 28 nm 0.9 V CMOS technology. The first charge-based transmitter realization consisted of a capacitive QDAC driving the PPA input through a 45 pF baseband capacitance. With a 10-bit DAC running at 128 MS/s, it demonstrates all the noise filtering capabilities of charge-based operation by achieving a noise floor notably 15 dB lower than of a TX using a conventional DAC (with the same number of bits and sampling frequency). At 45 MHz offset from a 1 GHz modulated carrier, it provides an out-of-band noise spectral density of −155 dBc/Hz, with ACLR1/2 of respectively −42 dB and −47 dB.

The same improved noise performance is also observed in the second implementation. In this case however, two different external baluns were used to validate the transmitter performance at both 900 MHz and 2.4 GHz. Achieving a peak power of 3.5 dBm from a 0.9 V supply, measurements with baseband frequencies ranging from 1 to 20 MHz show a maximum LO feedthrough and Image of −55 and −44 dBc, respectively. At 7 dB backoff, CIM3 at any baseband frequency is always below −50 dBc at both 900 MHz and 2.4 GHz, and as in the previous implementation the sampling aliases are shaped by a $\text{sinc}(x)^2$ transfer function, which corresponds to at least 20 dB of additional attenuation if compared to a conventional architecture. The ACLR 1/2 performance for a 20 MHz bandwidth signal is respectively −47 and −59 dB, with a measured EVM performance of 1.6 %. Finally, at 45 MHz offset the modulated noise power density of −159 dBc/Hz was measured at both LO frequencies, and what is very unique is the fact that even in backoff conditions, the noise performance is not significantly degraded, also thanks to the intrinsic noise filtering capabilities of the charge-based architecture. Therefore, with the achieved out-of-band noise performance and a core area consumption of only 0.22 mm^2, this architecture achieves what is—to the author's knowledge—the best out-of-band noise performance versus area consumption when compared to other similar works. ACLR and EVM performance are also among best.

In conclusion, valuable contributions were made to the current state-of-the-art by introducing an alternative charge-based transmitter architecture that can provide sensitive improvements in noise performance and area consumption. The achievement paves the way to small form-factor SAW-less fully digital multi-standard CMOS RF transmitter frontends, enabling advanced wireless communication systems including all the cellular standards 3G, 4G and beyond.

Bibliography

[Abi07] Abidi, A.A.: The path to the software-defined radio receiver. IEEE J. Solid-State Circuits **42**(5), 954–966 (2007). ISSN: 0018-9200. doi:10.1109/JSSC.2007.894307

[Abo99] Abo, A.M.: Design for reliability of low-voltage, switched-capacitor circuits. Ph.D. thesis, p. 145 (1999)

[Ala12] Alavi, M.S., Staszewski, R.B., de Vreede, L.C.N., Visweswaran, A., Long, J.R.: All-digital RF I/Q modulator. IEEE Trans. Microwave Theory Tech. **60**(11), 3513–3526 (2012). ISSN: 0018-9480. doi:10.1109/TMTT.2012.2211612

[Ala13] Alavi, M.S., Voicu, G., Staszewski, R.B., de Vreede, L.C.N., Long, J.R.: A 2x13-bit all-digital I/Q RF-DAC in 65-nm CMOS. In: 2013 IEEE Radio Frequency Integrated Circuits Symposium (RFIC), pp. 167–170. IEEE, Piscataway (2013). ISBN: 978-1-4673-6062-3. doi:10.1109/RFIC.2013.6569551

[Ala14] Alavi, M.S., Staszewski, R.B., de Vreede, L.C.N., Long, J.R.: A wideband 2x13-bit all-digital I/Q RF-DAC. IEEE Trans. Microwave Theory Tech. **62**(4), 732–752 (2014). ISSN: 0018-9480. doi:10.1109/TMTT.2014.2307876

[And14] Andrews, J.G., Buzzi, S., Choi, W., Hanly, S.V., Lozano, A., Soong, A.C.K., Zhang, J.C.: What will 5G be? IEEE J. Sel. Areas Commun. **32**(6), 1065–1082 (2014). ISSN: 0733-8716. doi:10.1109/JSAC.2014.2328098

[Aue11] Auer, G., Giannini, V., Desset, C., Godor, I., Skillermark, P., Olsson, M., Imran, M., Sabella, D., Gonzalez, M., Blume, O., Fehskc, A.: How much energy is needed to run a wireless network? IEEE Wirel. Commun. **18**(5), 40–49 (2011). ISSN: 1536-1284. doi:10.1109/MWC.2011.6056691

[Bag06] Bagheri, R., Mirzaei, A., Chehrazi, S., Heidari, M.E., Lee, M., Tang, M.M.W., Abidi, A.A.: An 800-MHz–6-GHz software-defined wireless receiver in 90-nm CMOS. IEEE J. Solid-State Circuits **41**(12), 2860–2876 (2006). ISSN: 0018-9200. doi:10.1109/JSSC.2006.884835

[Boo11] Boos, Z., Menkhoff, A., Kuttner, F., Schimper, M., Moreira, J., Geltinger, H., Gossmann, T., Pfann, P., Belitzer, A., Bauernfeind, T.: A fully digital multimode polar transmitter employing 17b RF DAC in 3G mode. In: 2011 IEEE International Solid-State Circuits Conference, pp. 376–378. IEEE, Piscataway (2011). ISBN: 978-1-61284-303-2. doi:10.1109/ISSCC.2011.5746361

[Bor13] Borremans, J., van Liempd, B., Martens, E., Cha, S., Craninckx, J.: A 0.9V low-power 0.4–6 GHz linear SDR receiver in 28 nm CMOS. In: 2013 Symposium on VLSI Circuits (2013)

P.E. Paro Filho et al., *Charge-based CMOS Digital RF Transmitters*, Analog Circuits and Signal Processing, DOI 10.1007/978-3-319-45787-1

[Cas09] Cassia, M., Hadjichristos, A., Kim, H.S., Ko, J.S., Yang, J., Lee, S.O., Sahota, G.: A low-power CMOS SAW-less quad band WCDMA/HSPA/HSPA+/1X/EGPRS transmitter. IEEE J. Solid-State Circuits **44**, 1897–1906 (2009). ISSN: 00189200. doi:10.1109/JSSC.2009.2020228

[Cha01] Chandran, N., Valenti, M.C.: Three generations of cellular wireless systems. IEEE Potentials **20**(1), 32–35 (2001). ISSN: 02786648. doi:10.1109/45.913210

[Chi10] Chironi, V., Debaillie, B., Baschirotto, A., Craninckx, J., Ingels, M.: An area efficient digital amplitude modulator in 90 nm CMOS. In: Proceedings of 2010 IEEE International Symposium on Circuits and Systems, pp. 2219–2222. IEEE, Piscataway (2010). ISBN: 978-1-4244-5308-5. doi:10.1109/ISCAS.2010.5537206

[Chi35] Chireix, H.: High power outphasing modulation. Proc. IRE **23**(11), 1370–1392 (1935). ISSN: 0096-8390. doi:10.1109/JRPROC.1935.227299

[Cho11] Chowdhury, D., Ye, L., Alon, E., Niknejad, A.M.: An efficient mixed-signal 2.4-GHz polar power amplifier in 65-nm CMOS technology. IEEE J. Solid-State Circuits **46**(8), 1796–1809 (2011). ISSN: 0018-9200. doi:10.1109/JSSC.2011.2155790

[Cod14] Codega, N., Rossi, P., Pirola, A., Liscidini, A., Castello, R.: A current-mode, low out-of-band noise LTE transmitter with a class-A/B power mixer. IEEE J. Solid-State Circuits **49**(7), 1627–1638 (2014). ISSN: 0018-9200. doi:10.1109/JSSC.2014.2315643

[Col14] Collados, M., Zhang, H., Tenbroek, B., Chang, H.-H.: A low-current digitally predistorted direct-conversion transmitter with 25% duty-cycle passive mixer. IEEE Trans. Microwave Theory Tech. **62**(4), 726–731 (2014). ISSN: 0018-9480. doi:10.1109/TMTT.2014.2309559

[Cra07] Craninckx, J., Liu, M., Hauspie, D., Giannini, V., Kim, T., Lee, J., Libois, M., Debaillie, D., Soens, C., Lngels, M., Baschirotto, A., Van Driessche, J., Van der Perre, L., Vanbekbergen, P.: A fully reconfigurable software-defined radio transceiver in 0.13 μm CMOS. In: 2007 IEEE International Solid-State Circuits Conference. Digest of Technical Papers, pp. 346–607. IEEE, Piscataway (2007). ISBN: 1-4244-0852-0. doi:10.1109/ISSCC.2007.373436

[Der09] Derudder, V., Bougard, B., Couvreur, A., Dewilde, A., Dupont, S., Folens, L., Hollevoet, L., Naessens, F., Novo, D., Raghavan, P., Schuster, T., Stinkens, K., Weijers, J.W., Van der Perre, L.: A 200 Mbps+ 2.14 nJ/b digital baseband multi processor system-on-chip for SDRs. In: 2009 Symposium on VLSI Circuits, pp. 292–293 (2009)

[Elo07] Eloranta, P., Seppinen, P., Kallioinen, S., Saarela, T., Parssinen, A.: A multimode transmitter in 0.13 um CMOS using direct-digital RF modulator. IEEE J. Solid-State Circuits **42**(12), 2774–2784 (2007). ISSN: 0018-9200. doi:10.1109/JSSC.2007.908749

[Fra09] Frappe, A., Flament, A., Stefanelli, B., Kaiser, A., Cathelin, A.: An all-digital RF signal generator using high-speed $\Sigma\Delta$ modulators. IEEE J. Solid-State Circuits **44**(10), 2722–2732 (2009). ISSN: 0018-9200. doi:10.1109/JSSC.2009.2028406

[Fuk12] Fukuda, S., Io, S.M.M., Hamashita, K., Nauta, B.: Direct-digital modulation (DIDIMO) transmitter with −156 dBc/Hz Rx-band noise using FIR structure. In: 2012 Proceedings of the ESSCIRC (ESSCIRC), pp. 53–56. IEEE, Piscataway (2012). ISBN: 978-1-4673-2213-3. doi:10.1109/ESSCIRC.2012.6341254

[Ful81] Fuller, R.B.: Critical Path, p. 471. St Martins Press, New York (1981). ISBN: 0-312-17491-8

[Gab11] Gaber, W.M., Wambacq, P., Craninckx, J., Ingels, M.: A CMOS IQ direct digital RF modulator with embedded RF FIR-based quantization noise filter. In: 2011 Proceedings of the ESSCIRC (ESSCIRC), pp. 139–142. IEEE, Piscataway (2011). ISBN: 978-1-4577-0703-2. doi:10.1109/ESSCIRC.2011.6044884

[Gia09] Giannini, V., Nuzzo, P., Soens, C., Vengattaramane, K., Ryckaert, J., Goffioul, M., Debaillie, B., Borremans, J., Van Driessche, J., Craninckx, J., Ingels, M.: A 2 mm^2 0.1–5 GHz software-defined radio receiver in 45-nm digital CMOS. IEEE J. Solid-State Circuits **44**(12), 3486–3498 (2009). ISSN: 0018-9200. doi:10.1109/JSSC.2009.2032585

[Gia11] Giannini, V., Ingels, M., Sano, T., Debaillie, B., Borremans, J., Craninckx, J.: A multiband LTE SAW-less modulator with −160 dBc/Hz RX-band noise in 40 nm LP CMOS. In: 2011 IEEE International Solid-State Circuits Conference, pp. 374–376. IEEE, Piscataway (2011). ISBN: 978-1-61284-303-2. doi:10.1109/ISSCC.2011.5746360

[GL10] 3GPP-LTE. In: Technical Specification Group Radio Access Network (Release 10) (2010)

[Glo03] Glossner, J., Iancu, D., Hokenek, E., Moudgill, M.: A software-defined communications baseband design. IEEE Commun. Mag. **41**(1), 120–128 (2003). ISSN: 0163-6804. doi:10.1109/MCOM.2003.1166669

[Gro07] Groe, J.: Polar transmitters for wireless communications. IEEE Commun. Mag. **45**(9), 58–63 (2007). ISSN: 0163-6804. doi:10.1109/MCOM.2007.4342857

[Har03] Harris, F.J., Dick, C., Rice, M.: Digital receivers and transmitters using polyphase filter banks for wireless communications. IEEE Trans. Microwave Theory Tech. **51**(4), 1395–1412 (2003). ISSN: 0018-9480. doi:10.1109/TMTT.2003.809176

[Har06] Harte, L.: Introduction to Mobile Telephone Systems, 2nd edn. Althos Publishing (2006)

[He09] He, X., van Sinderen, J.: A low-power low-EVM, SAW-less WCDMA transmitter using direct quadrature voltage modulation. IEEE J. Solid-State Circuits **44**(12), 3448–3458 (2009). ISSN: 0018-9200. doi:10.1109/JSSC.2009.2032495

[He10] He, X., van Sinderen, J., Rutten, R.: A 45 nm WCDMA transmitter using direct quadrature voltage modulator with high oversampling digital front-end. In: 2010 IEEE International Solid-State Circuits Conference - (ISSCC), pp. 62–63. IEEE, Piscataway (2010). ISBN: 978-1-4244-6033-5. doi:10.1109/ISSCC.2010.5434048

[Ing10] Ingels, M., Giannini, V., Borremans, J., Mandal, G., Debaillie, B., Van Wesemael, P., Sano, T., Yamamoto, T., Hauspie, D., Van Driessche, J., Craninckx, J.: A 5 mm^2 40 nm LP CMOS transceiver for a software-defined radio platform. IEEE J. Solid-State Circuits **45**(12), 2794–2806 (2010). ISSN: 0018-9200. doi:10.1109/JSSC.2010.2075210

[Ing13] Ingels, M., Furuta, Y., Zhang, X., Cha, S., Craninckx, J.: A multiband 40 nm CMOS LTE SAW-less modulator with −60 dBc C-IM3. In: 2013 IEEE International Solid-State Circuits Conference Digest of Technical Papers, pp. 338–339. IEEE, Piscataway (2013). ISBN: 978-1-4673-4516-3. doi:10.1109/ISSCC.2013.6487760

[Ing14] Ingels, M.: In: Baschirotto, A., Makinwa, K.A.A., Harpe, P. (eds.) Architectures for Digital Intensive Transmitters in Nanoscale CMOS. Springer, Cham (2014). ISBN: 978-3-319-01079-3. doi:10.1007/978-3-319-01080-9

[Jer07] Jerng, A., Sodini, C.G.: A wideband $\Delta\Sigma$ digital-RF modulator for high data rate transmitters. IEEE J. Solid-State Circuits **42**(8), 1710–1722 (2007). ISSN: 0018-9200. doi:10.1109/JSSC.2007.900255

[Jon07] Jones, C., Tenbroek, B., Fowers, P., Beghein, C., Strange, J., Beffa, F., Nalbantis, D.: Direct-conversion WCDMA transmitter with 163 dBc/Hz noise at 190 MHz offset. In: 2007 IEEE International Solid-State Circuits Conference. Digest of Technical Papers, pp. 336–607. IEEE, Piscataway (2007). ISBN: 1-4244-0852-0. doi:10.1109/ISSCC.2007.373431

[Kav08] Kavousian, A., Su, D.K., Hekmat, M., Shirvani, A., Wooley, B.A.: A digitally modulated polar CMOS power amplifier with a 20-MHz channel bandwidth. IEEE J. Solid-State Circuits **43**(10), 2251–2258 (2008). ISSN: 0018-9200. doi:10.1109/JSSC.2008.2004338

[Kay15] Kaymaksut, E., Reynaert, P.: Dual-mode CMOS doherty LTE power amplifier with symmetric hybrid transformer. IEEE J. Solid-State Circuits **50**(9), 1974–1987 (2015). ISSN: 0018-9200. doi:10.1109/JSSC.2015.2422819

[Kin13] King, A.L.S., Valença, A.M., Silva, A.C.O., Baczynski, T., Carvalho, M.R., Nardi, A.E.: Nomophobia: dependency on virtual environments or social phobia? Comput. Hum. Behav. **29**(1), 140–144 (2013). ISSN: 07475632. doi:10.1016/j.chb.2012.07.025

[Kun06] Kundert, K.: Simulating switched-capacitor filters with SpectreRF. In: The Designer's Guide Community, pp. 1–25 (2006); Electronic document accessed in Jan 2016 from www.designers-guide.org/analysis/sc-filters.pdf

[Lee03] Lee, T.H.: The Design of CMOS Radio-Frequency Integrated Circuits, 2nd edn. Cambridge University Press, Cambridge (2003)

[Lia13] Liang, P.C.P., Chien, G., Staszewski, R.B.: A 0.27 mm^2 13.5 dBm 2.4 GHz all-digital polar transmitter using 34%-efficiency class-D DPA in 40 nm CMOS. In: 2013 IEEE International Solid-State Circuits Conference Digest of Technical Papers, pp. 342–343. IEEE, Piscataway (2013). ISBN: 978-1-4673-4516-3. doi:10.1109/ISSCC.2013.6487762

[Lu13] Lu, C., Wang, H., Peng, C.H., Goel, A., Son, S., Liang, P., Niknejad, A., Hwang, H.C., Chien, G.: A 24.7 dBm all-digital RF transmitter for multimode broadband applications in 40 nm CMOS. In: 2013 IEEE International Solid-State Circuits Conference Digest of Technical Papers, pp. 332–333. IEEE, Piscataway (2013). ISBN: 978-1-4673-4516-3. doi:10.1109/ISSCC.2013.6487757

[Man16] Mangraviti, G., Khalaf, K., Shi, Q., Vaesen, K., Guermandi, D., Giannini, V., Brebels, S., Frazzica, F., Bourdoux, A., Soens, C., Van Thillo, W., Wambacq, P.: 13.5 A 4-antenna-path beamforming transceiver for 60 GHz multi-Gb/s communication in 28 nm CMOS. In: 2016 IEEE International Solid-State Circuits Conference (ISSCC), pp. 246–247. IEEE, Piscataway (2016). ISBN: 978-1-4673-9466-6. doi:10.1109/ISSCC.2016.7417999

[McC10] McCune, E.: Practical Digital Wireless Signals, vol. 1, p. 435. Cambridge University Press, Cambridge (2010). ISBN: 9788578110796. doi:10.1017/CBO9780511674648

[Meh09] Meher, P.K., Valls, J., Sridharan, K., Maharatna, K.: 50 years of CORDIC: algorithms, architectures, and applications. IEEE Trans. Circuits Syst. I: Regul. Pap. **56**(9), 1893–1907 (2009). ISSN: 1549-8328. doi:10.1109/TCSI.2009.2025803

[Meh10] Mehta, J., Staszewski, R.B., Eliezer, O., Rezeq, S., Waheed, K., Entezari, M., Feygin, G., Vemulapalli, S., Zoicas, V., Hung, C.-M., Barton, N., Bashir, I., Maggio, K., Frechette, M., Lee, M.-C., Wallberg, J., Cruise, P., Yanduru, N.: A 0.8 mm^2 all-digital SAW-less polar transmitter in 65 nm EDGE SoC. In: 2010 IEEE International Solid-State Circuits Conference - (ISSCC), pp. 58–59. IEEE, Piscataway (2010). ISBN: 978-1-4244-6033-5. doi:10.1109/ISSCC.2010.5434050

[Mir08] Mirzaei, A., Darabi, H.: A low-power WCDMA transmitter with an integrated notch filter. IEEE J. Solid-State Circuits **43**, 2868–2881 (2008). ISBN: 9781424420100. doi:10.1109/JSSC.2008.2005698

[Mir11a] Mirzaei, A., Murphy, D., Darabi, H.: Analysis of direct-conversion IQ transmitters with 25% duty-cycle passive mixers. IEEE Trans. Circuits Syst. I: Regul. Pap. **58**(10), 2318–2331 (2011). ISSN: 1549-8328. doi:10.1109/TCSI.2011.2142790

[Mir11b] Mirzaei, A., Darabi, H.: Analysis of imperfections on performance of 4-Phase passive-mixer-based high-Q bandpass filters in SAW-less receivers. IEEE Trans. Circuits Syst. I: Regul. Pap. **58**(5), 879–892 (2011). ISSN: 1549-8328. doi:10.1109/TCSI.2010.2089555

[Mit95] Mitola, J.: The software radio architecture. IEEE Commun. Mag. **33**(5), 26–38 (1995). ISSN: 01636804. doi:10.1109/35.393001

[Moo65] Moore, G.E.: Cramming more components onto integrated circuits. Electronics **38**(8), 114–117 (1965). ISSN: 0018-9219. doi:10.1109/jproc.1998.658762

[Ois14] Oishi, K., Yoshida, E., Sakai, Y., Takauchi, H., Kawano, Y., Shirai, N., Kano, H., Kudo, M., Murakami, T., Tamura, T., Kawai, S., Suto, K., Yamazaki, H., Mori, T.: A 1.95 GHz fully integrated envelope elimination and restoration CMOS power amplifier using timing alignment technique for WCDMA and LTE. IEEE J. Solid-State Circuits **49**(12), 2915–2924 (2014). ISSN: 0018-9200. doi:10.1109/JSSC.2014.2358554

[Oka11] Okada, K., Kousai, S.: Digitally-Assisted Analog and RF CMOS Circuit Design for Software-Defined Radio, edition published on 2011, Springer (2011)

[Oka14] Okada, K., Minami, R., Tsukui, Y., Kawai, S., Seo, Y., Sato, S., Kondo, S., Ueno, T., Takeuchi, Y., Yamaguchi, T., Musa, A., Wu, R., Miyahara, M., Matsuzawa, A.: 20.3 A 64-QAM 60 GHz CMOS transceiver with 4-channel bonding. In: 2014 IEEE International Solid-State Circuits Conference Digest of Technical Papers (ISSCC), pp. 346–347. IEEE, Piscataway (2014). ISBN: 978-1-4799-0920-9. doi:10.1109/ISSCC.2014.6757463

[Oli12] Oliaei, O., Kirschenmann, M., Newman, D., Hausmann, K., Xie, H., Rakers, P., Rahman, M., Gomez, M., Yu, C., Gilsdorf, B., Sakamoto, K.: A multiband multi-mode transmitter without driver amplifier. In: 2012 IEEE International Solid-State Circuits Conference, pp. 164–166. IEEE, Piscataway (2012). ISBN: 978-1-4673-0377-4. doi:10.1109/ISSCC.2012.6176960

[Oni13] Onizuka, K., Saigusa, S., Otaka, S.: A 1.8 GHz linear CMOS power amplifier with supply-path switching scheme for WCDMA/LTE applications. In: 2013 IEEE International Solid-State Circuits Conference Digest of Technical Papers, pp. 90–91. IEEE, Piscataway (2013). ISBN: 978-1-4673-4516-3. doi:10.1109/ISSCC.2013.6487650

[Opp97] Oppenheim, A.V., Willsky, A.S., Nawab, S.H.: Signals and Systems, Chap. 7, 2nd edn. (1997). ISBN: 0138147574

[Par09] Parikh, V.K., Balsara, P.T., Eliezer, O.E.: All digital-quadrature-modulator based wide-band wireless transmitters. IEEE Trans. Circuits Syst. I: Regul. Pap. **56**(11), 2487–2497 (2009). ISSN: 1549-8328. doi:10.1109/TCSI.2009.2015600

[Par15a] Filho, P.E.P., Wambacq, M.I.P., Craninckx, J.: A transmitter with 10 b 128 MS/S incremental-charge-based DAC achieving -155 dBc/Hz out-of-band noise. In: 2015 IEEE International Solid-State Circuits Conference (ISSCC) Digest of Technical Papers, pp. 164–165. IEEE, Piscataway (2015). ISBN: 978-1-4799-6223-5. doi:10.1109/ISSCC.2015.7062977

[Par15b] Filho, P.E.P., Ingels, M., Wambacq, P., Craninckx, J.: An incremental-charge-based digital transmitter with built-in filtering. IEEE J. Solid-State Circuits **50**(12), 3065–3076 (2015). ISSN: 0018-9200. doi:10.1109/JSSC.2015.2473680

[Par16] Filho, P.E.P., Ingels, M., Wambacq, P., Craninckx, J.: A 0.22 mm^2 CMOS resistive charge-based direct-launch digital transmitter with -159 dBc/Hz out-of-band noise. In: 2016 IEEE International Solid-State Circuits Conference (ISSCC), pp. 250–252 (2016). doi:10.1109/ISSCC.2016.7418001

[Per16] Perry, T.S.: Virtual reality goes social. IEEE Spectr. **53**(1), 56–57 (2016). ISSN: 0018-9235. doi:10.1109/MSPEC.2016.7367470

[Poz08] Pozsgay, A., Zounes, T., Hossain, R., Boulemnakher, M., Knopik, V., Grange, S.: A fully digital 65 nm CMOS transmitter for the 2.4-to-2.7 GHz WiFi/WiMAX bands using 5.4 GHz $\Delta\Sigma$ RF DACs. In: 2008 IEEE International Solid-State Circuits Conference - Digest of Technical Papers, pp. 360–619. IEEE, Piscataway (2008). ISBN: 978-1-4244-2010-0. doi:10.1109/ISSCC.2008.4523206

[Raz00] Razavi, B.: Design of Analog CMOS Integrated Circuits, 1st edn. McGraw-Hill (2000)

[Raz12] Razavi, B.: RF Microelectronics, 2nd edn. Pearson Education International, Upper Saddle River (2012). ISBN: 9780132839419

[Ros13] Rossi, P., Codega, N., Gerna, D., Liscidini, A., Ottini, D., He, Y., Pirola, A., Sacchi, E., Uehara, G., Yang, C., Castello, R.: An LTE transmitter using a class-A/B power mixer. In: 2013 IEEE International Solid-State Circuits Conference Digest of Technical Papers, pp. 340–341. IEEE, Piscataway (2013). ISBN: 978-1-4673-4516-3. doi:10.1109/ISSCC.2013.6487761

[Sha48] Shannon, C.E.: A mathematical theory of communication. Bell Syst. Tech. J. **27**, 379–423 (1928/1948). doi:10.1145/584091.584093

[Sow09] Sowlati, T., Agarwal, B., Cho, J., Obkircher, T., El Said, M., Vasa, J., Ramachandran, B., Kahrizi, M., Dagher, E., Vadkerti, M., Taskov, G., Seckin, U., Firouzkouhi, H., Saeidi, B., Akyol, H., Mahjoob, A., D'Souza, S., Guss, D., Shum, D., Badillo, D., Ron, I., Ching, D., Komaili, J., Loke, A., Pullela, R., Pehlivanoglu, E., Zarei, H., Tadjpour, S., Agahi, D., Rozenblit, D., Domino, W., Williams, G., Damavandi, N., Wloczysiak, S., Rajendra, S., Paff, A., Valencia, T.: Single-chip multiband WCDMA/HSDPA/HSUPA/EGPRS transceiver with diversity receiver and 3G DigRF interface without SAW filters in transmitter / 3G receiver paths. In: 2009 IEEE International Solid-State Circuits Conference Digest of Technical Papers, pp. 116–117, 117a. IEEE, Piscataway (2009). ISBN: 978-1-4244-3458-9. doi:10.1109/ISSCC.2009.4977335

[Sta04] Staszewski, R.B., Muhammad, K., Leipold, D., Wallberg, J.L., Fernando, C., Maggio, K., Staszewski, R., Jung, T., John, S., Sarda, V., Moreira-Tamayo, O., Mayega, V., Katz, R., Friedman, O., Eliezer, O.E., De-Obaldia, E., Balsara, P.T.: All-digital TX frequency synthesizer and discrete-time receiver for Blue-tooth radio in 130-nm CMOS. IEEE J. Solid-State Circuits **39**(12), 2278–2291 (2004). ISSN: 0018-9200. doi:10.1109/JSSC.2004.836345

[Sta05] Staszewski, R.B., Wallberg, J.L., Rezeq, S., Eliezer, O.E., Vemulapalli, S.K., Fernando, C., Maggio, K., Staszewski, R.B., Barton, N., Cruise, P., Entezari, M., Muhamma, K., Leipold, D.: All-digital PLL and transmitter for mobile phones. IEEE J. Solid-State Circuits **40**(12), 2469–2482 (2005). ISSN: 0018-9200. doi:10.1109/JSSC.2005.857417

[Ste12] Stewart, J.: Calculus, 7th edn. McMaster University, Toronto (2012). ISBN: 0538497815

[Van03] Van Den Bosch, A.: High resolution, high speed CMOS current-steering digital-to-analog converters. Ph.D. thesis (2003)

[Vid13] Vidojkovic, V., Szortyka, V., Khalaf, K., Mangraviti, G., Brebels, S., van Thillo, W., Vaesen, K., Parvais, B., Issakov, V., Libois, M., Matsuo, M., Long, J., Soens, C., Wambacq, P.: A low-power radio chipset in 40 nm LP CMOS with beamforming for 60 GHz high-data-rate wireless communication. In: 2013 IEEE International Solid-State Circuits Conference Digest of Technical Papers, pp. 236–237. IEEE, Piscataway (2013). ISBN: 978-1-4673-4516-3. doi:10.1109/ISSCC.2013.6487715

[Ye13a] Ye, L.: Design and analysis of digitally modulated transmitters for efficiency enhancement. Ph.D. thesis, p. 145 (2013)

[Ye13b] Ye, L., Chen, J., Kong, L., Alon, E., Niknejad, A.M.: Design considerations for a direct digitally modulated WLAN transmitter with integrated phase path and dynamic impedance modulation. IEEE J. Solid-State Circuits **48**(12), 3160–3177 (2013). ISSN: 0018-9200. doi:10.1109/JSSC.2013.2281142

[Yij03a] Zhou, Y., Yuan, J.: A 10-bit wide-band CMOS direct digital RF amplitude modulator. IEEE J. Solid-State Circuits **38**(7), 1182–1188 (2003). ISSN: 0018-9200. doi:10.1109/JSSC.2003.813290

[Yij03b] Zhou, Y., Yuan, J.: An 8-bit 100-Mhz CMOS linear interpolation DAC. IEEE J. Solid-State Circuits **38**(10), 1758–1761 (2003). ISSN: 0018-9200. doi:10.1109/JSSC.2003.817593

[Yin13] Yin, Y., Chi, B., Yu, Q., Liu, B., Wang, Z.: A 0.1–5 GHz SDR transmitter with dual-mode power amplifier and digital-assisted I/Q imbalance calibration in 65 nm CMOS. In: 2013 IEEE Asian Solid-State Circuits Conference (A-SSCC), pp. 205–208. IEEE, Piscataway (2013). ISBN: 978-1-4799-0280-4. doi:10.1109/ASSCC.2013.6691018

[Yon07] Yin, Y.-S., Gao, M.-L., Deng, H.-H., Liang, S.-Q., Liu, C.: A 14-bit 130-MSPS current-steering CMOS DAC with 2x FIR interpolation filter. In: 2007 7th International Conference on ASIC, pp. 703–706. IEEE, Piscataway (2007). ISBN: 978-1-4244-1131-3. doi:10.1109/ICASIC.2007.4415728

[Yoo12] Yoo, S.-M., Jann, B., Degani, O., Rudell, J.C., Sadhwani, R., Walling, J.S., Allstot, D.J.: A class-G dual-supply switched-capacitor power amplifier in 65 nm CMOS. In: 2012 IEEE Radio Frequency Integrated Circuits Symposium, pp. 233–236 (2012). ISBN: 978-1-4673-0416-0. doi:10.1109/RFIC.2012.6242271

[You11] Youssef, M., Zolfaghari, A., Darabi, H., Abidi, A.: A low-power wideband polar transmitter for 3G applications. In: 2011 IEEE International Solid-State Circuits Conference, pp. 378–380. IEEE, Piscataway (2011). ISBN: 978-1-61284-303-2. doi:10.1109/ISSCC.2011.5746362

Printed in the United States
By Bookmasters